INTRODUCTION TO LIBRARY RESEARCH IN GEOGRAPHY

An Instruction Manual and Short Bibliography

by

TOM L. MARTINSON

The Scarecrow Press, Inc.
Metuchen, N.J.   1972

Library of Congress Cataloging in Publication Data

Martinson, Tom L
    Introduction to library research in geography.

    Includes bibliographical references.
    1. Geographical research. 2. Reference books--
Geography. 3. Geography--Bibliography--Theory, methods,
etc. I. Title.
Z6001.M36   910'.7'2   72-2012
ISBN 0-8108-0495-6

PREFACE

This manual has been developed for use by students of geography who wish to expand their library research horizons. The roots of this work lie in the author's experience as a graduate student in geography at the University of Kansas and as a teacher of library research techniques in geography at the University of Colorado and Ball State University. This guide reflects the nature of library research in geography, and is designed to answer questions commonly asked by beginning graduate students in that field. Developed as a handbook for students of geography, this work may also prove useful to anyone who must provide answers to the broad range of research questions geographers ask.

The author acknowledges his debt to many individuals, both students and other colleagues, who contributed to the growth of this volume. Among those who provided inspiration and assistance are Thomas R. Smith of the University of Kansas, Anthony Shipps of Indiana University, Joseph S. Sabol of the University of Missouri - Kansas City, Dederick C. Ward of the University of Colorado, and Lynn S. Mullins of the American Geographical Society. Gratefully acknowledged is the assistance of the University of Colorado and Ball State University for making earlier, mimeographed editions of this work available to my students and other interested persons. The work, advice, and criticism of students welded this volume into a useful product; any shortcomings and errors within it are my responsibility.

iii

CONTENTS

Page

Preface   iii

Introduction   7

Chapter

I.   The Stratified Flow Chart for Compiling
     Subject Bibliographies in Geography   16

II. Printed Library Catalogs   30

III. Guides to Serials   39

IV. References to Government Publications   44

V.   Statistics Sources   55

VI. References to Theses and Dissertations   69

VII. Subject Bibliographies in Geography   71
     Comprehensive Guides to Geography   71
     Printed Library Catalogs in Geography   72
     Guides to Serials in Geography   72
     References to Government Publications
       in Geography   73
     Statistics Sources for Geography   73
     References to Theses and Dissertations
       in Geography   75
     Guides to Geography Specialties   75

Summary and Notes for Further Research   133

Author Index   137

INTRODUCTION

Perhaps the prime hallmark of a scholar is his ability to propose meaningful questions and then conduct a comprehensive search for answers. This manual is designed to assist geography students in finding answers to their research problems.

In recent years, the search for answers in geography as well as in other fields has been complicated by a vast increase in the amount of information available, a comparable increase in the rate of information generated, and an increase in the number and influence of different media presenting the information. [1]

Because most answers provided by geographers in the course of their research are still found in the printed word and accompanying maps, this manual is concerned primarily with the search for these printed materials. The device most valuable for conducting this search for published research is bibliography. [2]

Bibliographies can be viewed as agents to control ("store") and find ("retrieve") published research. They can be consulted before and during research in order to discover the ways other geographers have proposed and answered questions as well as new methods they have developed to collect and manipulate data. [3]

Bibliography is both a technique and a finished product. The technique of compiling a bibliography incorporates many skills, including those which test the researcher's ability to be accurate, consistent and straightforward. The completed bibliography is a refined and useful tool which allows subsequent researchers to analyze the evidence from which conclusions are drawn. [4]

Just as bibliography is both a tool and a product, this manual can be used as an aid to compiling a subject bibliography and as a short guide to geographical materials. The emphasis is on recent United States reference works. [5]

7

This manual is organized to reflect the interdiscipli-
nary realm of the geographer. Six major subject subdivisions
are presented in order, according to their degree of compre-
hensiveness. Major, comprehensive works are presented in
the sections on Library Catalogs and Guides to Serials. Two
types of secondary reference works of special interest to
geographers, Government Publications and Statistics Sources,
are included next. The following section on References to
Theses and Dissertations includes guides to original research
in many disciplines, including geography. The section on
Subject Bibliographies in Geography, which constitutes a
large portion of this manual, includes both comprehensive
and secondary reference works. Aside from the organization by sections, this manual also contains organization of
reference sources using a relatively new technique, the flow
chart. Several flow charts are contained in this manual;
each is designed to be helpful in organizing specific bibliographical sources for easy library search. Notes which
amplify and extend points of interest are found at the end
of every section.

Notes

1.   For comments on the increasing complexity of science
and technology and the resulting complexity in reference
bibliography, see the following representative sources:

> Kent, Allen. Specialized information centers. New York,
> Spartan, 1965. (A1)

> Becker, Joseph and Hayes, Robert M. Information
> storage and retrieval. New York, Wiley, 1965. (A2)

> Schultheiss, Louis A., Culbertson, Don and Heilinger,
> Edward M. Advanced data processing in the univer-
> sity library. New York, Scarecrow, 1962. (A3)

> Licklider, J. C. R. Libraries of the future. Cambridge,
> M. I. T. Press, 1965. (A4)

> Smailes, A. A. "The future of scientific and techno-
> logical publications." Aslib Proceedings, 22:48-54,
> 1970. (A5)

2.   The term "bibliography" refers to the technique of compilation and the final listing of sources. This manual is con-

cerned primarily with the compilation of bibliographies.   A
second definition of "bibliography" divides the field into
descriptive and systematic elements.   Descriptive bibliography refers to the physical makeup of books, while systematic bibliography is concerned with the assembling of information contained in books.   This manual is concerned
primarily with systematic bibliography.   There are three
types of systematic bibliography: 1) author-title compilations; 2) bibliographies with critical abstracts; and 3) annotated bibliographies.   Each of these types of bibliographies may be subject oriented; that is, fashioned to meet
the needs of a relatively limited number of persons with a
common research interest.   This manual is concerned primarily with the compilation of subject bibliographies, whether
author-title, critically abstracted or annotated.   For further
information on the definition of bibliography, see

> Wynar, Bohdan S.   Introduction to bibliography and
> reference work; a guide to materials and sources.
> 3rd ed. rev. and enl.   Denver, Libraries Unlimited,
> 1966. (A6)

The following list includes some general bibliographical
guides which, although international and comprehensive, contain many subject bibliographies and guides useful to geographers.

> Winchell, Constance M.   Guide to reference works.   8th
> ed.   Chicago, American Library Association, 1967.
> (A7)

> Sheehy, Eugene P.   Guide to reference works.   8th ed.
> First supplement, 1965-1966.   Chicago, American
> Library Association, 1968. (A8)

> _____. "Selected reference books of 1968-69." College
> and Research Libraries, 31:109-117, March 1970.
> (A9)

> _____. "Selected reference books of 1969-70." College
> and Research Libraries, 31:269-279, July 1970. (A10)

> Walford, A. J. (ed.)   Guide to reference materials.
> 2nd ed.   London, Library Association, 1966- .   In
> progress. (A11)

> Besterman, Theodore.   A world bibliography of bibliog-

raphies and of bibliographical catalogues, calendars,
abstracts, digests, indexes and the like.   4th ed.
Geneva, Societas Bibliographica, 1965-1966.   5 vols.
(A12)

Bibliographic index; a cumulative bibliography of bibliographies. 1937- . New York, Wilson, 1938- . Semiannual with annual cumulations. (A13)

Shores, Louis. Basic references; an introduction to
materials and methods. Chicago, American Library
Association, 1954. (A14)

Collison, Robert L. Bibliographies, subject and national; a guide to their contents, arrangement and use.
2nd ed. rev. and enl. New York, Hafner, 1962.
(A15)

_____. Bibliographical services throughout the world,
1950-1959. Paris, UNESCO, 1961. (A16)

Avicenne, Paul. Bibliographical services throughout
the world, 1960-1964. Paris, UNESCO, 1969.
(Supplemented monthly in the UNESCO publication
titled Bibliography, Documentation, Technology.)
(A17)

      The search for entries in a subject bibliography need
not be a complicated process.   There are relatively few
steps which a researcher must take in planning his bibliography.   These steps are illustrated in the "General Flow
Chart for Bibliography Search" (see page 11).   As the flow
chart indicates, there are two alternative means of beginning
bibliographic search: consulting library accessions lists,
handbooks and guides or comprehensive, secondary and
specialized reference sources.   As the shaded pattern indicates, this manual is concerned primarily with introducing
the latter sources.   The remainder of the flow chart consists of "question boxes" and "choice lines" which direct
the researcher on a course of discovery through a bibliography research problem.   If the researcher responds with
a series of "no" answers to the location questions asked in
the boxes, his search for bibliography entries ends in
failure.   If he responds with a series of "yes" answers,
he begins to compile a bibliography from sources he knows
and can find.   The most important of the comprehensive,
secondary and specialized sources for building subject bib-

GENERAL FLOW CHART
FOR BIBLIOGRAPHY SEARCH

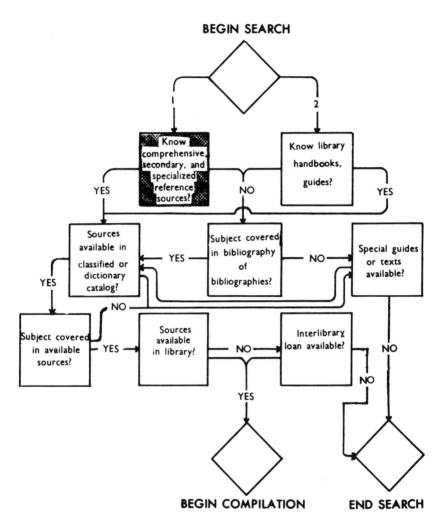

(See Stratified Flow Chart for Bibliography Compilation)

 **This manual is concerned with this step
in bibliography search and compilation**

liographies in geography are listed in the "Stratified Flow
Chart for Bibliography Compilation" found in Chapter 1 of
this manual.

3. The organization of a bibliography should reflect the objectives of the researcher and the nature of his research
problem.   It should be extensive enough to cover all pertinent phases of his topic and intensive enough to treat especially important points in some detail.
At times sources take on added significance if properly
placed within a structural framework.   A related but seem-
ingly unimportant fragment of information could become the
keystone of an argument developed later during the writing
of a report.
Structuring a bibliography helps focus attention on
the development of major themes or processes of the topic
under study.   Understanding these themes aids in the organization of the final product of research and may open
the door to further investigation.
Organization of the bibliography should suggest itself
from the nature of the problem investigated.   A bibliography
should be flexible enough to admit future additions as well
as retrospective and current sources.   If a researcher
wishes to structure a bibliography to serve as an outline
for larger and more complex projects, or to serve as a
basis for an expanding interest in geography, it may be
best to adopt the general schemes of large classification
systems.
There are effective methods of indexing small bibliographies on note cards, including the use of edge-notched
cards in which subjects are represented by holes punched
in appropriate places on a regular 3 x 5 or 4 x 6 inch card.
A sharp, slender object pushed through a combination of
notches and holes would retrieve all cards dealing with that
subject.   For further information see:

Bourne, C. P.   Methods of information handling.   New
York, Wiley, 1965.   (A18)

Snyder, David E.   "Peripheral punch cards in geographic research. "   Professional Geographer, 12:
13-15, November 1960.   (A19)

One way to organize a bibliography to save time in
subsequent bibliographic compilation is to consult a classified library catalog (which contains entries listed by Library
of Congress or Dewey Decimal systems, as in library shelf-

lists) rather than a dictionary library catalog (which contains entries listed by author, subject or title, as in the public card catalog). It is necessary to consult guides to the classified catalog before proceeding to the shelflists. The two classification systems used as guides in most libraries and therefore the most valuable in organizing bibliographies on broad geographical themes are presented in the following works:

Dewey, Melvil. Dewey decimal classification and relative index. 17th ed. New York, Forest Press, Inc. of Lake Placid Club Education Foundation, 1965. 2 vols. (A20)

United States. Library of Congress. Subject Cataloging Division. Classification class G; geography, anthropology, folklore, manners and customs, recreation. 3rd ed. Washington, Government Printing Office, 1954. (A21)

After the Dewey and Library of Congress guides have been consulted to obtain the classification numbers for the subject under study, the student may use the library shelflist to compile a bibliography of the holdings of his local library on his chosen topic. This bibliography may provide a framework for a larger, more comprehensive compilation.
Some libraries may deny use of their shelflist to the beginning researcher. Moreover, use of the shelflist requires some experience in bibliography compilation.
The flow chart presented with these notes may be valuable to the beginning researcher who needs experience in bibliography compilation because it provides a framework for a step-by-step search for information. See the flow charts in the following chapters as well. For further information on flow charts in reference search, consult such works as the following:

Bunge, Charles A. "Charting the reference query." RQ, 8:245-250, Summer 1969. (A22)

Winings, J. W. "A data structure for information retrieval." Journal of the American Society for Information Science, 21:145-148, March-April 1970. (A23)

Martinson, Tom L. "A simple stratified flow chart

for compiling subject bibliographies in geography. "
Special Libraries Association Geography and Map
Division Bulletin, 82:32-41, December 1970. (A24)

4. Part of the technique of bibliography is the compilation
of personal bibliography cards. The elements of bibliography
cards considered absolutely necessary are: 1) the author's
full name, inverted; 2) the title and subtitle; 3) the edition,
if not the first; and 4) the imprint, which includes the place
of publication, the publisher and the date of publication.

Consistency in type of information presented and the
order of presentation, and uniformity in size of bibliography
cards are also absolutely essential. It is necessary to de-
velop early a card "style" which suits the needs of the research and the researcher. The value of consistency be-
comes more and more apparent as the bibliography grows in
size.

A bibliography which does not contain cards with accurate, complete entries is of little or no value. Experienced researchers have acquired the habit of making note
cards instead of merely listing sources in the course of research. A collection of sources entered properly on note
cards is a highly functional and valuable reference to use in
further research. For further information on bibliographic
technique, see the following bibliography:

Kinney, Mary R. Bibliographical style manuals; a
guide to their use in documentation and research.
Chicago, Association of College and Research Libraries, 1953. (A25)

This bibliographical style manual is very useful:

McCrum, Blanche P. , and Jones, Helen D. Bibliographical procedures and style; a manual for bibliographers in the Library of Congress. Washington,
Government Printing Office, 1966. (A26)

Among the most commonly used manuals illustrating ac-
ceptable research paper style are:

Turabian, Kate L. A manual for writers of term
papers, theses, and dissertations. rev. ed. Chicago,
University of Chicago Press, 1965. (A27)

_____. Student's guide for writing college papers.
Chicago, University of Chicago Press, 1963. (A28)

5. Reference librarians have developed careers in the dissemination of research information and therefore are inestimably valuable resource persons. The following is a highly selective list of readings on the history and nature of reference work.

Cole, Dorothy E. "Some characteristics of reference work." College and Research Libraries, 7:45-51, January 1946. (A29)

Harris, K. G. "Reference service today and tomorrow: objectives, practices, needs and trends." Journal of Education for Librarianship, 3:175-187, Winter, 1963. (A30)

Kaplan, Louis. "Reference service in university and special libraries since 1900." College and Research Libraries, 19:217-220, May 1958. (A31)

Wilson, Logan. "Library roles in American higher education." College and Research Libraries, 31: 96-102, March 1970. (A32)

Chapter I

THE STRATIFIED FLOW CHART FOR COMPILING
SUBJECT BIBLIOGRAPHIES IN GEOGRAPHY

The much-heralded information explosion has extended
to all academic disciplines, geography included. The profusion of information has made it advisable for geographers
wishing to keep abreast of current research to use increas-
ingly complex and sophisticated techniques of bibliographic
control such as those provided by computers.

Geographers need not turn to computers to solve all
their problems of bibliographic control. There exist a wide
variety of simple manual methods of bibliography compilation
which can be used by scholars with limited time and purpose.
Some of these techniques have been inspired by computer
programming.[1]

The purpose of this chapter is to introduce briefly one
standard and straightforward manual technique which geographers may use to deal with a profusion of information.
This technique is the flow chart, a means of arranging information simply and clearly, which is commonly used in
programming. The flow chart presented in this brief report is designed to aid in the compilation of subject bibliographies.[2]

Flow Charts as Keys to Information Arrangement

Bibliographies are agents used to control ("store")
and find ("retrieve") information. Viewed in this light,
bibliographies are mental computers with a vast capacity
to simplify and shorten information storage and retrieval
problems. Geographers may find that these mental computers can be programmed effectively with the help of flow
charts such as the one presented here.

Geographers may find that flow charts, among other
information arrangement techniques, are more useful than

the standard bibliographical guides librarians commonly produce to lead students into geography as a subject field. Many
librarians, trained to perceive geography as little more than
the encyclopedic accumulation of place names or isolated spatial facts, can provide geographers with little guidance in
bibliography search and compilation.[3]

Flow charts are of great potential value in organizing
sources used by geographers in bibliography compilation. In
general, flow charts illustrate a series of steps required for
the solution of a particular problem. The flow chart introduced here illustrates the series of sources a researcher
with limited time and purpose would investigate first in preparation of a specialized subject bibliography in geography.

The flow chart presented here is not useful for compiling or arranging facts such as place names, terms or other information likely to be found in dictionaries or gazetteers.
On the other hand, the flow chart is useful in gathering entries for bibliographies developed on specific subjects or
problems.

Bibliographical flow charts are particularly useful
techniques for geographers because many geographers are
accustomed to implementing mental concepts to deal with
complex abstractions. The region, a term common in geographical literature, has been defined as a mental tool to
deal with the complexity of areas. Statistical models now
frequent in geographical literature are mental constructs as
well, although designed to analyze many more variables and
the more subtle interrelations between them.

The highly simplified flow chart presented in this paper is one which leads the geography researcher on an ab-
breviated tour of the most essential reference works necessary for compiling subject bibliographies on specific geographical research problems. The present flow chart is designed to emphasize contemporary United States publications
and English-language works. The more experienced scholar
could construct a more detailed and complete flow chart to
suit the bibliographical needs of his specialized research
problems.

Description of the Flow Chart Diagram

Each box in the flow chart diagram introduced in this

chapter represents one category of references. In most
cases, each box represents an individual reference source,
although some boxes represent a reference series. Consulting the reference-boxes and gleaning their pertinent entries
are steps toward compiling a subject bibliography in geography.

The flow chart diagram employs two visual devices to
indicate the character of the references represented by the
boxes. Darkly shaded boxes refer to general reference
works, while lightly shaded boxes refer to geographical reference works. The thick black lines separate groups of
boxes into categories according to whether they are comprehensive works (such as library catalogues and guides to serials), secondary works (such as guides to government documents, statistics sources or guides to theses and dissertations), or highly specialized geographical works designed for
compiling highly specialized subject bibliographies. The
thick black lines as well as the shading patterns give the
flow chart its stratified appearance.

The number in each box refers to the individual works
or series of sources listed on the pages following the flow
chart diagram. Each of the twenty-three boxes represents
one general reference work or series or one basic geographical reference work or series. The references in the boxes
should be consulted in order, in the sequence indicated by
the numbers and the direction of the arrows.

There are two alternative routes through the flow
chart diagram. A researcher may wish to consult only the
general works in the darkly shaded boxes or only the geographical reference works in the lightly shaded boxes. In
either case, the references should be consulted in the sequence indicated by the numbers and arrows.

The researcher who consults each source (with the
exception of the first and the last) in numerical order will
encounter geography references first and general references
second. The bibliography compiler using the sources in this
order can save time in two ways. First, he can familiarize
himself with a broad category of reference works using a
more familiar source before proceeding on to the more complex general sources. Second, he may find that the geographical bibliographies contain enough entires for his personal subject bibliography and the general reference works
need not be consulted in every case. The broad categories

of references in the flow chart diagram as divided by the thick black lines are found on the pages identified as "References to Accompany Stratified Flow Chart. "

The first and last entries in the flow chart diagram are unshaded to draw attention to their special value. The first entry, Winchell's Guide to Reference Works, is perhaps the most widely used comprehensive reference work in the United States. Among the sections of Winchell's Guide which are of special value to geographers are Bibliography (AA), especially pages 2-10; Periodicals (AF), pages 132-150; Government Publications (AH), pages 155-163; Statistics (CG), pages 372-387; Dissertations (AI), pages 163-167; Earth Sciences (EE), pages 560-570; and Geography (CK), pages 441-461.

The last entry in the flow chart diagram is unshaded because it is reserved for the many highly specialized topical, regional and technical bibliographies in geography which are available to scholars with highly specialized research problems. Samples of these specialized bibliographies are at the end of the references accompanying the flow chart.

The flow chart presented here indicates only the most essential English-language sources for the compilation of a subject bibliography in geography. It is highly selective, and most useful to geographers with well-defined research problems but little time to review all the available literature and compile a bibliography of pertinent references. The flow chart thus may be viewed as a time saving manual program for a mental computer, and a prototype for more complex information storage and retrieval systems in an age of information abundance.

STRATIFIED FLOW CHART
FOR BIBLIOGRAPHY COMPILATION

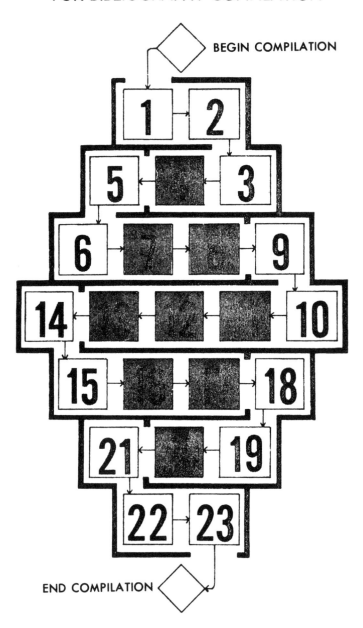

REFERENCES TO ACCOMPANY STRATIFIED FLOW CHART

Comprehensive Reference Sources

1. Winchell, Constance M. Guide to reference works.
   8th ed. Chicago, American Library Association,
   1967. (A7)

Library Catalogues

2. Wright, John K., and Platt, Elizabeth T. Aids to geographical research; bibliographies, periodicals, atlases,
   gazetteers and other reference books. 2nd ed. rev.
   New York, Columbia University Press for the American Geographical Society, 1947. (B1)

3. American Geographical Society. Library. Research
   catalogue. Boston, G. K. Hall, 1962. 15 vols. (B2)

   _____ . _____ . Current geographical publications; additions to the Research catalogue of the American Geographical Society. New York, American Geographical
   Society, vol. 1- . 1938- . 10 issues yearly. (B3)

4. United States. Library of Congress. A catalog of books
   represented by Library of Congress printed cards, issued to July 31, 1942. Ann Arbor, Michigan, Edwards, 1942-46. 167 vols. (B4)

   _____ . _____ . Supplement: Cards issued August 1,
   1942-December 31, 1947. Ann Arbor, Edwards, 1948.
   42 vols. (B5)

   _____ . _____ . Library of Congress author catalog: a
   cumulative list of works represented by Library of
   Congress printed cards, 1948-1952. Ann Arbor, Edwards. 1953. 24 vols. (B6)

   _____ . _____ . The National Union Catalog: A cumulative author list representing Library of Congress
   printed cards and titles reported by other American
   libraries, 1953-1957. Ann Arbor, Edwards, 1958.
   26 vols. (B7)

_____. _____. Library of Congress author catalog: A cumulative author list representing Library of Congress printed cards and titles reported by other American libraries, 1958-1962. New York, Rowman and Littlefield, 1963. 54 vols. (B8)

_____. _____. The National Union Catalog: A cumulative author list representing Library of Congress printed cards and titles reported by other American libraries, 1963-1967. Ann Arbor, Edwards, 1969. 59 vols. (B9)

_____. _____. The National Union Catalog: A cumulative author list representing Library of Congress printed cards and titles reported by other American libraries, 1968. Washington, Library of Congress, 1969. 12 vols. (B10)

_____. _____. The National Union Catalog: A cumulative author list representing Library of Congress printed cards and titles reported by other American libraries, 1969. Washington, Library of Congress, 1970. 13 vols. (B11)

_____. _____. Library of Congress catalog. Books: subjects, 1950-1954; a cumulative list of works represented by Library of Congress printed cards. Ann Arbor, Edwards, 1955. 20 vols. (B12)

_____. _____. Library of Congress catalog. Books: subjects, 1955-1959; a cumulative list of works represented by Library of Congress printed cards. Paterson, New Jersey, Pageant Books, 1960. 22 vols. (B13)

Guides to Serials

5. Harris, Chauncy D. Annotated world list of selected current geographical serials in English: including an appendix of major serials in other languages with regular supplementary of partial basic use of English. 2nd ed. rev. and enl. Chicago, University of Chicago, 1964. (University of Chicago, Department of Geography, Research paper no. 96.) (B14)

6. Harris, Chauncy D., and Fellmann, Jerome D. Inter-

national list of geographical serials. Chicago, Uni-
versity of Chicago, 1960. (University of Chicago,
Department of Geography, Research paper no. 63.)
(B15)

7. Ulrich's international periodicals directory; a classified
guide to current periodicals, foreign and domestic.
13th ed. New York, Bowker, 1969-70. 2 vols.
(B16)

8. Union list of serials in libraries of the United States
and Canada. 3rd ed. by Edna B. Titus. New York,
Wilson, 1965. 5 vols. (B17)

New serial titles, 1950-1960; supplement to the Union
list of serials, 3rd ed. A union list of serials com-
mencing publication after December 31, 1949- .
Prepared under the sponsorship of the Joint Committee on the Union List of Serials. Washington, Library of Congress, 1961. 2 vols. (B18)

Guides to Government Publications

9. Vinge, Clarence L., and Vinge, A. G. United States
government publications for research and teaching in
geography and related social and natural sciences.
Totowa, New Jersey, Littlefield, Adams, 1967.
(B19)

10. Felland, Nordis. "United Nations publications useful to
geographers." The Professional Geographer, 10:11-
13, July 1958. (B20)

11. Andriot, John L. Guide to United States government
serials and periodicals. McLean, Virginia, Documents Index, 1966. 4 vols. (B21)

12. Schmeckebier, Laurence F., and Eastin, Roy B. Government publications and their use. 2nd ed. rev.
Washington, The Brookings Institution, 1969. (B22)

13. Wilcox, Jerome K. Manual on the use of state publications. Chicago, American Library Association,
1940. (B23)

Statistics Sources

14. Anderson, Marc. A working bibliography of mathematical geography. Ann Arbor, Michigan, University of Michigan, Department of Geography, 1963. (Michigan Inter-University Community of Mathematical Geographers Discussion paper no. 2.) (B24)

15. Berry, Brian J. L., and Pred, Allan R. Central place studies: a bibliography of theory and applications. Philadelphia, Regional Science Research Institute, 1961. (Bibliography series no. 1.) (B25)

    Barnum, H. G., Kasperson, R., and Kiuchi, S. Central place studies ... Supplement, 1965. Philadelphia, Regional Science Research Institute, 1965. (B26)

16. Lancaster, Henry O. Bibliography of statistical bibliographies. Edinburgh, Oliver and Boyd for the International Statistical Institute, 1968. (B27)

17. Andriot, John L. Guide to United States government statistics. 3rd ed. rev. and enl. Arlington, Virginia, Documents Index, 1961. (B28)

Guides to Theses and Dissertations

18. Whittlesey, Derwent. Dissertations in geography accepted by universities in the United States for the degree of Ph.D. as of May, 1935. Annals of the Association of American Geographers, 25:211-237, December 1935. (B29)

19. Browning, Clyde E. A bibliography of dissertations in geography: 1901 to 1969. Chapel Hill, N.C., University of North Carolina, Department of Geography, 1970. (Studies in geography no. 1.) (B30)

20. Dissertation abstracts international: abstracts of dissertations and monographs in microform. Ann Arbor, Michigan, University Microfilms, 1952- . vol. 12- . Monthly. (B31)

Specialized Geography Bibliographies

21. Harris, Chauncy D. Bibliographies and reference works
for research in geography. Chicago, University of
Chicago, Department of Geography, 1967. (B32)

22. Lewthwaite, Gordon R., Price, Edward T., Winters,
Harold A. A geographical bibliography for American
college libraries; a revision of A Basic geographical
library, a selected and annotated book list for American Geographers, 1970. (Commission on College
Geography publication no. 9.) (B33)

Selected Bibliographies in Topical, Regional,
and Technical Fields of Geography

23. Ward, Dederick C. Geologic reference sources.
Metuchen, N.J.: Scarecrow Press, 1972. (B34)

United States. Weather Bureau. Selective guide to
published climatic data sources. Washington, Government Printing Office, 1969. (Key to meteorological records documentation no. 4.11.) (B35)

Water Resources Council. Hydrology and Sedimentation
Committees. Annotated bibliography on hydrology
and sedimentation, 1963-1965, United States and
Canada. Washington, Water Resources Council,
1969. (Joint hydrology-sedimentation bulletin no.
9.) Supplement to previous bibliographies on hydrology and sedimentation that were prepared in
cooperation with or under the auspices of the Subcommittee on Hydrology and Sedimentation, Inter-
Agency Committee on Water Resources. Published
through arrangements made by the Soil Conservation
Service acting in behalf of several participating governmental agencies. (B36)

American Meteorological Society. Meteorological and
geoastrophysical abstracts. Boston, American Meteorological Society, 1950- . Monthly. (B37)

National Research Council. Committee on Oceanography.
Oceanographic information sources; a staff report.

Washington, National Academy of Sciences-National
Research Council, 1970. (B38)

Blake, Sidney F. Geographical guide to floras of the
world. New York, Hafner, 1963. Reprint of the
1942 edition of the United States Department of Agriculture, Miscellaneous publication No. 401. (B39)

Smith, Roger C. Guide to the literature of the zoological sciences. 6th ed. Minneapolis, Burgess, 1962.
(B40)

Bibliography of soil science, fertilizers and general
agronomy, 1959-1962. Farnham Royal, England,
Commonwealth Agricultural Bureaux, 1964. (B41)

Olsson, Gunnar. Distance and human interaction: a
review and bibliography. Philadelphia, Regional
Science Research Institute, 1965. (Regional Science Research Institute Bibliography series no. 2.)
(B42)

Bestor, George C., and Jones, Holway R. City plan-
ning: a basic bibliography of sources and trends.
Rev. ed. Sacramento, California Council of Civil
Engineers and Land Surveyors, 1966. (B43)

Branch, Melville C. Comprehensive urban planning:
a selective annotated bibliography with related materials. Beverly Hills, California, Sage, 1970.
(B44)

Holleb, Doris B. Social and economic information for
urban planning. Chicago, Center for Urban Studies
of the University of Chicago, 1969. 2 vols. (B45)

National Association of Home Builders of the United
States. Basic texts and reference books on housing
and construction; a selected, annotated bibliography.
Washington, National Association of Home Builders,
1956. (B46)

Zelinsky, Wilbur. A bibliographic guide to population
geography. Chicago, University of Chicago, 1962.
(University of Chicago, Department of Geography,
Research paper no. 80.) (B47)

Stevens, Benjamin H., and Brackett, Carolyn A. Industrial location: a review and annotated bibliography of theoretical, empirical and case studies.
Philadelphia, Regional Science Research Institute,
1967. (Regional Science Research Institute Bibliography series no. 3.) (B48)

Wang, Jen Yu, and Barger, Gerald L. (eds. and
comps.) Bibliography of agricultural meteorology.
Madison, University of Wisconsin Press, 1962.
(B49)

Siddall, William R. Transportation geography: a bibliography. Manhattan, Kansas, Kansas State University, 1967. (Kansas State University Library
Bibliography series no. 1.) (B50)

Food and Agriculture Organization of the United Nations.
Bibliography on land tenure. New York, United
Nations, 1955. (B51)

ReQua, Eloise G., and Statham, Jane. The developing
nations: a guide to information sources concerning
their economic, political, technical and social problems. Detroit, Gale Research Company, 1965.
(Management information guide series no. 5.) (B52)

Harmon, Robert B. Political science; a bibliographical
guide to the literature. New York, Scarecrow,
1965. (B53) (Supplements, 1968 and 1972)

Peltier, Louis C. (ed.) Bibliography of military geography. Washington, Association of American Geographers, 1962. (B54)

McManis, Douglas R. Historical geography of the
United States: a bibliography--excluding Alaska
and Hawaii. Ypsilanti, Michigan, Eastern Michigan
University, 1965. (B55)

Selected Regional Bibliographies

International Geographical Union. Special Commission
on the Humid Tropics. A select annotated bibliography of the humid tropics. Comp. by Theo L.
Hills. Montreal, McGill University, 1960. (B56)

American Universities Field Staff. A select bibliogra-
    phy: Asia, Africa, Eastern Europe, Latin America.
    New York, American Universities Field Staff, 1960
    (and supplements). (B57)

Berry, Brian J. L., and Hankins, Thomas D. A bib-
    liographic guide to the economic regions of the
    United States. Chicago, University of Chicago,
    1963. (University of Chicago, Department of Geog-
    raphy Research paper no. 87.) (B58)

Gropp, Arthur E. (comp.) A bibliography of Latin
    American bibliographies. Washington, Pan Ameri-
    can Union, 1968. (B59)

United States. Library of Congress. Reference Divi-
    sion. Soviet geography, a bibliography. Ed. by
    Nicholas R. Rodionoff. Washington, Government
    Printing Office, 1951. 2 vols. (B60)

Twentieth Century Fund. Survey of tropical Africa;
    select annotated bibliography of tropical Africa.
    Ed. by C. Daryll Forde. New York, International
    African Institute, 1956. (B61)

Hall, Robert B., and Noh, Toshio. Japanese geogra-
    phy: a guide to Japanese reference and research
    materials. Ann Arbor, Michigan, University of
    Michigan Press, 1956. (University of Michigan
    Center for Japanese Studies Bibliographic series
    no. 6.) (B62)

Pelzer, Karl J. Selected bibliography on the geography
    of Southeast Asia. New Haven, Yale University,
    1949-1956. 3 vols. (B63)

Selected Bibliographies on Geographical Techniques

Porter, Philip W. A bibliography of statistical cartog-
    raphy. Minneapolis, University of Minnesota Book-
    store, 1964. (B64)

American Geographical Society. Map Department. In-
    dex to maps in books and periodicals. Boston, G.
    K. Hall, 1967. (B65)

Gywer, Joseph A., and Waldron, Vincent G. Photo
interpretation techniques; a bibliography. Washington, Library of Congress, 1956. (B66)

Walters, Robert L. Radar bibliography for geoscientists. Lawrence, Kansas, University of Kansas,
1968. (CRES Report no. 61-30, Remote Sensing
Laboratory.) (B67)

Gabler, Robert E. (ed.) A handbook for geography
teachers. Normal, Illinois, National Council for
Geographic Education, 1966. (Geographic education
series no. 6.) (B68)

Notes

1. Among the books on programming which explain flow
chart techniques is:

McCameron, Fritz. FORTRAN logic and programming.
Homewood, Ill., Irwin, 1968. (B69)

Two sources on the manual retrieval of information using
flow chart methods are:

Bunge, Charles A. "Charting the reference query."
RQ, 8:245-250, Summer 1969. (A22)

Dale, Doris C. "The manual retrieval of government
publications." The Bookmark, 29:106-110, December
1969. (B70)

2. A subject bibliography, for purposes of this paper, is
defined as an organized series of bibliographical citations
compiled on a specific topic in the course of research.

3. Of course, of great value to geographers are those rare
librarians trained in earth science, area studies and map
classification and cataloguing. Two organizations dedicated
to the training of these individuals are the Geoscience Information Society and the Geography and Map Division of the
Special Libraries Association.

Chapter II

PRINTED LIBRARY CATALOGS

A union library catalog is one example of the comprehensive reference works used in systematic subject bibliography. These bibliographies of library holdings are considered equivalent to other universal reference works such as
bibliographies of bibliographies, guides to book selection,
guides to reference books, national and trade bibliographies,
book review indexes and guides to serials as primary sources
for entries in a personal, working bibliography.[1]

Union library catalogs, although they are only one of
the comprehensive reference works available to geography
bibliographers, are extremely important in bibliographic
search. The catalogs of large national libraries, such as
the Library of Congress, are very useful for bibliographical
research because they often contain entries on every book
published in their respective countries. The catalogs of the
Library of Congress, the British Museum, the Bibliothèque
Nationale and the Deutscher Gesamtkatalog, for example, are
the most comprehensive single records of the publications of
the United States, Great Britain, France and Germany. Very
few countries in the world have compiled such union catalogs,
although many have attempted to do so.[2]

The author catalogs produced by the United States Library of Congress are the only ones considered here. They
are most readily available to American scholars, and they
may be considered illustrative of those produced or attempted
by other countries.

Printed library catalogs, such as the Library of Congress author catalogs, are valuable to the beginning and experienced researcher for a number of reasons. For example, they describe book contents and editions, give information on authors, supply the location of at least one copy of
the books listed and supply verification of titles for interli-
brary loan purposes. The Library of Congress author series is especially valuable to geographers engaged in area

30

research, for countries are considered authors and therefore censuses and other national data sources on the shelves
of the Library of Congress are listed in its author series.
The Library of Congress subject catalogs, not considered
here, contain sources arranged by subject. The subject
headings refer to a wide range of concepts, including those
related to geography.[3]

Use of the Stratified Flow Chart
in Consulting Library of Congress Author Catalogs

The stratified flow chart, shown in Chapter I, contains boxes which represent categories of references. The
complete stratified flow chart is subdivided by thick black
lines indicating category of reference work (comprehensive,
secondary or specialized) and by shading patterns indicating
general or geographical sources. The thick black lines as
well as the shading patterns give the flow chart its stratified
appearance.

Each box represents one category of references important in developing a subject bibliography in geography.
Consulting each box (reference source) in order enables the
researcher to gather entries for bibliographies developed on
specific subjects or problems. The simplified flow chart
leads the geography researcher on a brief tour of the most
essential reference works used in compiling subject bibliographies in his field.

Library of Congress Section of the Stratified Flow Chart

Boxes three and four in the stratified flow chart represent library catalogs. Box three refers to the American
Geographical Society's Research Catalogue and Current Geographical Publications.[2] Box four, representing the several
United States Library of Congress catalogs, is examined
here.

Enlargement of the Library of Congress Catalog box
(entry number four in the enlarged portion of the stratified
flow chart) reveals an author catalog "microseries" within
it. The "microseries" diagram is composed of circles
which represent the Library of Congress author catalogs.
The numbers and lower case letters in the diagram refer
to the names of the sources and indicate the order in which

they should be consulted during bibliographical search. The
sources are listed on the pages identified as "References to
Accompany Library of Congress Author Catalog Microseries"
following the flow chart.

The Library of Congress Author Catalogs are among
the most important comprehensive reference works available
to American scholars, but one drawback to the value of these
printed library catalogs is their variation in name and degree
of coverage. The following suggestions for the use of the
Library of Congress Author Catalogs may be helpful to the
beginning researcher.

Prior to 1969, the search for the works of a particular author in the Library of Congress catalogs included an
examination of 1) the Main Catalog, except for books published later than July, 1942; 2) the Supplement; 3) the suc-
ceeding volumes of the Library of Congress National Union
Catalog under its various names and formats; 4) the quarterly
supplements; and 5) the monthly supplements. Two cumulative catalogs begun in 1969, however, may be used to simplify the search through the Library of Congress catalogs.
These cumulative catalogs are 1) the National Union Catalog
Pre-1956 Imprints and 2) the Library of Congress and National Union Catalog Author Lists, 1942-1962: A Master
Cumulation. Full citations for these sources are provided
on the pages following the flow chart.

As a result of the addition of the new cumulative catalogs begun in 1969, there are two alternative routes through
the Library of Congress author catalog "microseries" diagram. The researcher chooses one of the two routes depending on the nature of his research problem and the length
of time available for bibliography compilation.

Route one consists of the National Union Catalog Pre-
1956 Imprints, the Library of Congress and National Union
Catalog Author Lists, 1942-1962: A Master Cumulation, and
the National Union Catalog supplements and cumulations from
1963 to the present. Use of this route is time-saving, for
the researcher must consult comparatively few sources. This
route may be more valuable for research in that it consists
of more recently-compiled catalogs and probably contains
more complete listings of sources. Both the Pre-1956 Imprints and the Master Cumulation are still in process, however, and their alphabets are not complete as yet.

Route two includes the Catalog of Books Represented by Library of Congress Printed Cards Issued to July 31, 1942 (sometimes referred to as the Main Catalog) followed by all its supplements and cumulations to the present. This route is suggested for those who are more interested in studying the nature and organization of the Library of Congress catalogs or those whose research carries them beyond the present alphabetical limitations of Pre-1956 Imprints and the Master Cumulation.

Researchers who consult any of the Library of Congress author catalogs contained in the two alternative routes can expect to reap rich bibliographical rewards. As Constance Winchell has said:

> Because of the immensity of the collections, the excellence of the cataloging and the full bibliographical descriptions, the Catalog of the Library of Congress has been for many years an invaluable work in any library and indispensable in those where research is done.[4]

LIBRARY OF CONGRESS CATALOG
'MICROSERIES'
IN THE STRATIFIED FLOW CHART

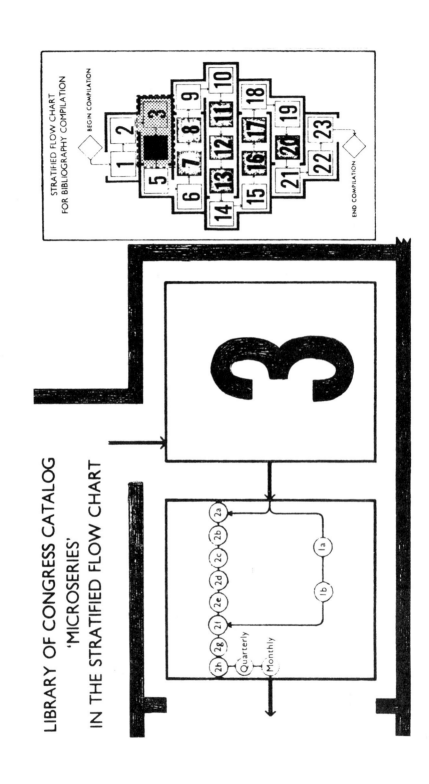

STRATIFIED FLOW CHART
FOR BIBLIOGRAPHY COMPILATION

BEGIN COMPILATION

END COMPILATION

REFERENCES TO ACCOMPANY LIBRARY OF CONGRESS
AUTHOR CATALOG "MICROSERIES"

1a. United States. Library of Congress. The National
     Union Catalog; Pre-1956 imprints. A cumulative
     author list representing Library of Congress printed
     cards and titles reported by other American librar-
     ies. London, Mansell Information/Publishing, Ltd.,
     1968- . In process. (C1)

1b.   .   . Library of Congress and National Union
     Catalog author lists, 1942-1962: A master cumula-
     tion. A cumulative author list representing entries
     in the Library of Congress--National Union Catalog
     supplements to the Catalog of books represented by
     Library of Congress printed cards. Detroit, Gale
     Research Co., 1969- . In process. (C2)

2a.   .   . A catalog of books represented by Li-
     brary of Congress printed cards, issued to July 31,
     1942. Ann Arbor, Edwards, 1942-1946. 167 vols.
     (B4).

2b.   .   . Supplement: Cards issued August 1,
     1942-December 31, 1947. Ann Arbor, Edwards,
     1948. 42 vols. (B5)

2c.   .   . Library of Congress author catalog: a
     cumulative list of works represented by Library of
     Congress printed cards, 1948-1952. Ann Arbor, Ed-
     wards. 1953. 24 vols. (B6)

2d.   .   . The National Union Catalog: A cumula-
     tive author list representing Library of Congress
     printed cards and titles reported by other American
     libraries, 1953-1957. Ann Arbor, Edwards, 1958.
     26 vols. (B7)

2e.   .   . Library of Congress author catalog: A
     cumulative author list representing Library of Con-
     gress printed cards and titles reported by other
     American libraries, 1958-1962. New York, Rowman
     and Littlefield, 1963. 54 vols. (B8)

2f.   .   . The National Union Catalog: A cumula-
     tive author list representing Library of Congress

printed cards and titles reported by other American
libraries, 1963-1967. Ann Arbor, Edwards, 1969.
59 vols. (B9)

2g. _____ . _____ . The National Union Catalog: A cumulative author list representing Library of Congress
printed cards and titles reported by other American
libraries, 1968. Washington, Library of Congress,
1969. 12 vols. (B10)

2h. _____ . _____ . The National Union Catalog: A cumulative author list representing Library of Congress
printed cards and titles reported by other American
libraries, 1969. Washington, Library of Congress,
1970. 13 vols. (B11)

Notes

1. The comprehensive bibliographies not included in this
manual are listed in one general reference guide already referred to, namely:

Winchell, Constance M. Guide to reference works. 8th
ed. Chicago, American Library Association, 1967.
(A7)

as well as its supplements which, together, are considered
the indispensable reference librarian's tool.
American trade bibliographies are one category of
reference works extremely valuable for persons wishing to
keep abreast of current book publishing in geography. See,
for example, the following sources:

Publishers' trade list annual. New York, R. R. Bowker,
1873- . Annual. (C3)

Books in print; an author-title index to the Publishers'
trade list annual. New York, R. R. Bowker, 1948- .
Annual. (C4)

Subject guide to books in print: an index to the Publishers' trade list annual. New York, R. R. Bowker,
1957- . Annual. (C5)

Publishers' Weekly: the American book trade journal.
New York, R. R. Bowker, 1872- . Weekly. (C6)

BPR: American book publishing record ... as cataloged by the Library of Congress and annotated by Publishers' Weekly. New York, R. R. Bowker, 1960- . Monthly. (C7)

Cumulative book index: a world list of books in the English language. New York, H. W. Wilson, 1898- . Monthly except August. (C8)

Forthcoming books. New York, R. R. Bowker, 1966- . Bimonthly. (C9)

Of special interest to those entering the elementary or secondary teaching field is:

Textbooks in print, including teaching materials, 1956- . New York, R. R. Bowker, 1956- . Annual. (C10)

Vertical file index: subject and title index to selected pamphlet material. New York, H. W. Wilson, 1935- . Monthly except August. (C11)

2. Among the important guides to national bibliographies are the following:

Courtney, William P. A register of national bibliography; with a selection of the chief bibliographical books and articles printed in other countries. London, Constable, 1905-12. 3 vols. (C12)

Larsen, Knud. National Bibliographical services, their creation and operation ... Paris, UNESCO, 1953. (UNESCO bibliographical handbooks, no. 1.) (C13)

United States. Library of Congress. General Reference and Bibliography Division. Current national bibliographies, comp. by Helen F. Conover. Washington, Government Printing Office, 1955. (C14)

3. The books listed in the National union catalog and its predecessors are not necessarily also found in the Library of Congress catalog: books: subjects. Books: subjects is a supplement to the National Union Catalog, and contains publications catalogued by the Library of Congress and 500 cooperating libraries beginning in 1945. The entries contained in the main catalog (United States Library of Congress: A catalog of books represented by Library of Congress

38   Library Research in Geography

printed cards issued to July 31, 1942) and many of those in
the first supplement to the main catalog are then not included
in Books: subjects.
    The various supplements to the Library of Congress
catalogs do not necessarily contain entries for books published in the time period the supplement covers. The first
supplement to the Library of Congress main catalog, for
example, contained entries for all books cataloged by the
Library of Congress from August 1, 1942, to December 31,
1947, regardless of the date of publication of the books.
    The introduction to each Library of Congress catalog
series is invaluable as a guide to the structure and purpose
of the series. For further information on the organization
and utility of the Library of Congress catalogs, consult the
following representative reviews.

    Bishop, William Warner. "Union catalogs." Library
        Quarterly, 7:36-49, January 1937. (C15)

    David, Charles W. "The National Union Catalog, 1952-
        1955 Imprints--a publication in early prospect." Col-
        lege and Research Libraries, 21:275-76, July 1960.
        (C16)

    Gourde, Leo. "Evolution of the National Union Catalog."
        Catholic Library World, 31:272-277, 315-316, Febru-
        ary 1960. (C17)

    Harman, Marian. "The National Union Catalog, A Re-
        view." Library Resources and Technical Services,
        2:209-215, Summer 1958. (C18)

    Miller, Marvin A. "Union Catalogs--national, regional
        and local." Library Journal, 66:59-62, January 1,
        1941. (C19)

4. Winchell, Constance M. Guide to reference works. 8th
        ed. Chicago, American Library Association, 1967,
        pp. 8. (A7)

Chapter III

GUIDES TO SERIALS

A second type of comprehensive reference works used in systematic subject bibliography is guides to serials. These guides are exceedingly valuable bibliographic tools because serials lead researchers to the most recent information available on their topics, provide information on subjects which may not be reported elsewhere, offer valid descriptions and analyses of events as they occurred and often contain current bibliographies in their fields.[1]

There are at least three types of guides to serials: bibliographies, union lists and indexes. Bibliographies of serials list general information on their titles and organization. Union lists are publications which indicate the location of serials in national or regional libraries. Serials indexes are similar to library catalogs in that they describe contents and editions, give information on publishers and supply the location of at least one set of the periodicals listed.

Serial bibliographies and union lists of serials are instruments used to choose the serials which should be investigated for appropriate entries in the researcher's personal bibliography. Although geographers are aware of the major serial publications within their discipline, many other serials contain articles of interest and need for geographic research.

Serial Bibliographies

Many national and regional serial bibliographies have been published. The most comprehensive lists have been compiled for France, Great Britain, the Soviet Union and the United States. Among the most important of the current bibliographies of serials in the United States are the following:

Ayer, N. W., and Son (firm). N. W. Ayer and Son's Directory of newspapers and periodicals. Philadelphia, Ayer, 1880- . vol. 1. Annual. (D1)

Editor and Publisher. International yearbook number
for 1920- . New York, Editor and Publisher, 1920- .
Annual. (D2)

Ulrich's international periodicals directory: a classified
guide to a selected list of current periodicals, foreign
and domestic. 13th ed. New York, R. R. Bowker,
1969-70. 2 vols. (B16)

Ulrich's international periodicals directory, Annual supplement, 1- . New York, R. R. Bowker, 1966- .
(D3)

Union Lists of Serials

Union lists of serials are valuable because they locate
sets of periodicals and therefore may assist the researcher
in obtaining photocopies of articles he may need in his studies, but the geographer is unlikely to use them in the normal
course of bibliographical search. There are three union
lists of serials which are consulted regularly in the United
States:

United States. Library of Congress. General Reference
and Bibliography Division. Union lists of serials:
a bibliography. Washington, Government Printing Office, 1964. (D4)

Union list of serials in libraries of the United States and
Canada. 3rd ed. by Edna B. Titus. New York, Wilson, 1965. 5 vols. (B17)

New serials titles, 1950-1960; Supplement to the Union
list of serials, 3rd ed. A union list of serials com-
mencing publication after December 31, 1949- . Prepared under the sponsorship of the Joint Committee on
the Union List of Serials. Washington, Library of
Congress, 1961. 2 vols. Continued by monthly issues and annual cumulations. (B18)

Indexes to Serials

After using serial bibliographies and union lists of
serials to choose appropriate serial publications, the researcher must find individual articles within those serial

publications to build his bibliography. These individual articles can be located using a serial index. Indexes to serials contain citations to serial articles arranged by topic or subject subdivisions. This arrangement usually is referred to as dictionary cataloging.

Indexes to serials are valuable guides because they provide important information about the serials themselves. Many serials indexes have a long history of publication as well as full dictionary cataloging of all articles, including subject, author and title entries. Common in the libraries of the United States are the following indexes. Most important for geography research are the last four entries in the series.

Poole's index to periodical literature, 1802-81. rev. ed. Boston, Houghton, 1891. 2 vols. (D7)

Nineteenth Century readers' guide to periodical literature, 1890-1899, with supplementary indexing, 1900-1922. New York, Wilson, 1944. 2 vols. (D8)

Readers' guide to periodical literature, 1900- . New York, Wilson, 1905- . vol. 1- . Monthly. (D9)

Social sciences and humanities index. New York. H. W. Wilson, 1916- . (Formerly International index to periodicals.) Quarterly with annual cumulations. (D10)

Public Affairs Information Service. Bulletin. New York, Public Affairs Information Service, 1915- . Weekly, with cumulations five times yearly and annually. (D11)

Applied science and technology index. New York, Wilson, 1913- . Monthly, with quarterly and annual cumulations. (D12)

Biological abstracts; reporting the world's biological research literature. Philadelphia, BioScience's Information Service of Biological Abstracts, 1926- . Twice monthly. (D13)

Use of Guides to Serials

Use of guides to serials in building a subject bibliog-

raphy is relatively easy. If the searcher is unfamiliar with
the serial publications of geography, his first step would be
to consult the latest Ulrich's international periodicals directory section on geography. This section in Ulrich's 1969
edition contains 87 entries, each referring to a specific geography serial. Within the reference to each geography serial
is the name of the publication which indexes the serial. Examination of the entries on English-language geography serials in Ulrich's directory reveals that the indexes serving the
most English-language geography serials are the Public Affairs Information Service Bulletin (PAIS), Social Sciences and
Humanities Index, and Biological Abstracts. The researcher
interested in finding the most geography articles in the short-
est length of time would then consult the above three indexes.
These indexes are arranged by subject and area subdivisions.

    The procedure involved in using guides to serials may
be illustrated by an example. Suppose the beginning researcher in the geography of Central America consults Ulrich's directory and finds that the Annals of the Association of American Geographers is a major English-language geography serial and therefore a likely source of geographic information on
Central America. The entry in Ulrich's directory reveals
that the Annals is indexed in Biological Abstracts and the Social Sciences and Humanities Index. The researcher then
finds these two indexes, turns to their sections on Central
America, and observes if there are any entries from the Annals. If so, they become part of the researcher's bibliography. Of course, these indexes contain entries from many
other serials besides the Annals so they are rich sources of
information on the serial literature of geography and other
disciplines.

Notes

1. Newspapers are in some ways superior to serials in
their ability to provide immediacy to bibliographies. No general index to newspapers is available. The following specialized indexes may be of some value in obtaining entries for
subject bibliographies:

        New York Times index, vol. 1- . 1933- . New York,
            The Times, 1913- . Semimonthly. (D14)

        Christian Science Monitor. Subject index of the Christian
            Science Monitor, vol. 1- . January 1960- . Boston,

Christian Science Monitor, 1960- . Monthly. (D15)

2. The Readers' Guide has been praised for the uniformity
of its entries, the use of consistent subject headings instead
of catchwords, the full information contained in its references
and its use of cumulation which keeps the index up to date.
The Readers' Guide to Periodical Literature, however,
indexes only popular periodicals likely to be found in the
average American public library, not periodicals designed
for geographic research. Serials of peripheral importance
to geographic research indexed by the Readers' Guide include
the Annals of the American Academy of Political and Social
Science, the Congressional Digest, Current History, Focus,
Foreign Affairs, the National Geographic, Scientific American, Science, the UNESCO Courier and Weatherwise.
For further information on serial indexes, consult the
following sources:

Askling, J. "Confusion worse confounded: how to eval-
uate an index." California Librarian, 13:90-92, December 1951. (D16)

_____. "What is an index?" California Librarian, 12:
159-60, 177, March 1951. (D17)

Clapp, V. W. "Indexing and abstracting: recent, past
and lines of future development." College and Research Libraries, 11:197-206, July 1950. (D18)

Huff, W. H. "Indexing, abstracting and translation
services." Library Trends, 10:427-447. January
1962. (D19)

"Indexing of book reviews in periodicals." Dartmouth
College Library Bulletin, 5:12-16. April 1962. (D20)

Jacobstein, J. M. "Indexes and indexing: a selected
bibliography of periodical articles." Library Journal,
83:1357-1358, May 1, 1958. (D21)

McColvin, L. R. "Purpose of indexing." Indexer, 1:31-
35, Summer 1958. (D22)

Steiner-Prag, E. F. "Indexes and indexing: a selected
bibliography of books and pamphlets." Library Journal, 83:1356-1357, May 1958. (D23)

Chapter IV

REFERENCES TO GOVERNMENT PUBLICATIONS

This section introduces references referred to as
"secondary" in the Introduction to the manual. Guides, indexes and bibliographies of government publications are considered here as secondary reference works because their
coverage is not designed to be as comprehensive or universal as library catalogs or guides to serials.[1]

Just as governments are complex in their organization,
the publications produced by governments are varied in character. Government publications are definitely not limited to
description or analysis of government processes; many are
valuable to physical and social scientists as well as experts
in other fields.[2]

Because most international organizations and individual
countries maintain publishing houses, government publications
account for a high proportion of world production. Guides to
this massive publishing output then become exceedingly valuable.

Guides to Multi-National Publications

There are a number of references to international
publications. One of the best guides is:

Brown, Everett S. Manual of government publications:
United States and foreign. New York, Appleton, 1950.
(E1)

One of the best bibliographies is:

Childs, James B. Government document bibliography in
the United States and elsewhere. 3rd ed. Washington, Government Printing Office, 1942. (E2)

and one of the best union lists is:

Gregory, Winifred (ed.) List of the serial publications of foreign governments, 1815-1931. New York, Wilson, 1932. (E3)

These references have been supplemented by more recent articles in library serials.

Guides to the production of several related countries are available in various works, including the following:

United States. Library of Congress. African Section. Official publications of British East Africa. Washington, Government Printing Office, 1960-1963. 4 vols. (E4)

_____. _____. Official publications of French Equatorial Africa, French Cameroons, and Togo, 1946-1958. Washington, Government Printing Office. 1964. (E5)

_____. _____. Madagascar and adjacent islands; a guide to official publications. Washington, Library of Congress, 1965. (E6)

_____. _____. General Reference and Bibliography Division. Official publication of French West Africa, 1946-1958. Washington, Government Printing Office, 1960. (E7)

Hill, Roscoe R. The National archives of Latin America. Cambridge, Harvard University Press, 1945. (E8)

United States. Library of Congress. A guide to the official publications of the other American republics. Washington, Government Printing Office, 1945-1949. 19 vols. (E9)

The international organizations which publish materials most valuable to the geographer are the United Nations and its predecessor, the League of Nations. Most experts agree that the most useful guides to these publications are the following:

Aufricht, Hans. Guide to League of Nations publications: a bibliographical survey of the work of the League, 1920-1947. New York, Columbia University Press, 1951. (E10)

Brimmer, Brenda (and others). A guide to the use of
    United Nations documents. Dobbs Ferry, New York,
    Oceana, 1962. (E11)

United Nations. Dag Hammarskjold Library. Checklist
    of United Nations documents, 1946-1959. New York,
    United Nations, 1949-1953. (E12)

United Nations. Dag Hammarskjold Library. Documents
    Index Unit. United Nations documents index. January,
    1950- . New York, United Nations, 1950- . vol. 1- .
    Monthly. (E12)

Guides to United States Government Publications

The Government Printing Office of the United States
is one of the most prolific in the world, and it produces a
great variety of publications. A recent issue of Selected
United States Government Publications, for example, lists
such publications as "The Weather Bureau and Water Management," "Coastal Warning Facilities Charts," "Thailand Area
Handbook," "Education and Outdoor Recreation," and "Selected Readings in Techniques of Stereotaxic Brain Surgery."

A number of guides, bibliographies and indexes have
been developed to make this vast range of information available to researchers.[2] Four useful guides are:

Boyd, Anne M. United States government publications.
    3rd ed. rev. by Rae E. Rips. New York, H. W.
    Wilson, 1949. (E13)

Hirshberg, Herbert S. and Melinat, Carl H. Subject
    guide to United States government publications. Chi-
    cago, American Library Association, 1947. (E14)

Leidy, W. Phillip. A popular guide to government pub-
    lications, New York, Columbia University Press, 1968.
    (E15)

Schmeckebier, Lawrence F. and Eastin, Roy B. Gov-
    ernment publications and their use. 2nd ed. rev.
    Washington, Brookings Institution, 1969. (B22)

Two useful bibliographies are:

Andriot, John L. Guide to United States government
serials and periodicals. McLean, Virginia, Documents Index, 1966. 4 vols. (B21)

Body, Alexander C. Annotated bibliography of bibliographies on selected government publications and supplementary guides to the Superintendent of Documents
classification system. Ann Arbor, Michigan, Edwards, 1967. (E16)

and the most important general catalogs and indexes are:[3]

Poore, Benjamin P. A descriptive catalog of the government publications of the United States, September
5, 1774- . March 4, 1881. Washington, Government Printing Office, 1885. (E17)

Ames, John G. Comprehensive Index to the publications
of the United States government, 1881-1893. Washington, Government Printing Office, 1905. 2 vols.
(E18)

United States. Superintendent of Documents. Checklist
of United States public documents, 1789-1909. Washington, Government Printing Office, 1911. (E19)

_____. _____. Catalog of the public documents of Con-
gress and of all departments of government of the
United States for the period March 4, 1893-December
31, 1940. Washington, Government Printing Office,
1896-1945. Vols. 1-25. (E20)

_____. _____. Monthly catalog of United States govern-
ment publications. 1895- . Washington, Government
Printing Office, 1895- . Monthly. (E21)

_____. _____. Selected list of United States government
publications. Washington, Government Printing Office,
1928- . Biweekly. (E22)

At times researchers are interested in information
which is presented in legislative debates in Congress. One
guide to this information is:

United States. Congress. Congressional record: containing the proceedings and debates of the 43rd Con-
gress- . March 4, 1873- . Washington, Government

Printing Office, 1873- . Issued daily; bound volumes
for each session. (E23)

Often congressional reports are known better by popular name than by bill number. Individuals interested in trac-
ing legislation known by popular name will find the following
source invaluable.

United States. Library of Congress. Serials Division.
Popular names of United States government reports;
a catalog. Washington, Government Printing Office,
1966. (E24)

The researcher interested in the structure of the
United States government should consult:

United States government organization manual. 1935- .
Washington, Government Printing Office. 1935- .
Annual. (E25)

Guides to State Government Publications

State publications may also prove valuable in some
geographic research. Valuable general guides to state publications are:[4]

Council of State Governments. Book of the states, vol.
1- . Chicago, Council of State Governments, 1935- .
Biennial. (E26)

_____. Index to council of state government publications.
Chicago, Council of State Governments, 1960.
(E27)

Edwards, Richard A. Index digest of state constitutions,
2nd ed. New York, Legislative Drafting Research
Fund of Columbia University, 1959. (E28)

United States. Federal Works Agency. Work Projects
Administration. Catalogue, WPA writers' program
publications; The American Guide Series, The American Life Series. Washington, Government Printing
Office, 1941. (E29)

Graves, William B. (and others). American state government and administration, a state-by-state bibliog-

raphy of significant general and special works. Chicago, Council of State Governments, 1949. (E30)

United States. Library of Congress. Photoduplication Service. A guide to the microfilm collection of early state records. ed. by Lillian A. Hamrick, Washington, Library of Congress, 1950. (E31)

_____. _____. A guide to the microfilm collection of early state records: supplement. ed. by William S. Jenkins. Washington, Library of Congress, 1951. (E32)

Some commonly-used guides to state publications are the following:

Wilcox, Jerome K. Manual on the use of state publications. Chicago, American Library Association, 1940. (B23)

Lloyd, Gwendolyn. "The status of state documents bibliography," Library Quarterly, 18:192-195. July, 1948. (E33)

National Association of State Libraries. Public Documents Clearing House Committee. Check-list of legislative journals of the states of the United States of America. Providence, Rhode Island, Oxford Press, 1938. (E34)

Pullen, William R. A check list of legislative journals issued since 1937 by the states of the United States of America. Chicago, American Library Association, 1955. (E35)

United States. Library of Congress. Processing Department. Monthly check-list of state publications, 1 vol. 1910- . Washington, Government Printing Office, 1910- . Monthly. (E36)

Two early guides to the publications of several states are the following:

Bowker, R. R. A provisional list of the official publications of the several states of the United States from their organizations ... New York, Publisher's Weekly, 1899-1908. To 1899 for New England states. 1902

for North Central states. 1905 for Western states.
1908 for Southern states. (E37)

Carnegie Institution. Index of economic material in
documents of the United States. Comp. by Alelaide
Hasse. Washington, Carnegie Institution, 1902-1916.
(Thirteen states only.) (E38)

Individual states have also produced references to
their own publications. See, for example:

Press, Charles and Williams, Oliver. State manuals,
blue books, and election results. Berkeley, Institute
of Governmental Studies, University of California,
1962. (E39)

Tompkins, Dorothy C. State government and administration; a bibliography. Berkeley, Bureau of Public Administration, University of California, 1954. (E40)

Guides to Local Government Publications

Local governments have also produced references to
their own publications. Four bibliographies on the affairs
and statistics of local governments are:

Hodgson, James G. Official publications of American
countries, a union list. Fort Collins, Colorado State
College, 1937. (E41)

Government Affairs Foundation, Inc. Metropolitan com-
munities; a bibliography, with special emphasis upon
government and politics. Chicago, Public Administration Service, 1957. (E42)

_____. Supplement, 1955-1957. Chicago, Public Ad-
ministration Service, 1960. (E43)

Municipal year book, 1934- ; the authoritative résumé
of activities and statistical data of American cities.
Chicago, International City Managers' Association,
1934- . Annual. (E44)

Use of References to Government Publications

Most students interested in locating recent United States government publications necessary for their research will find that the Monthly Catalog is their most important source of information. The other sources listed in the preceding sections are more specialized and therefore less valuable as general sources of recent United States publications. The user of the Monthly Catalog should refer to its index first in searching for entries. The index lists authors (usually government agencies) and titles alphabetically under topic subheadings. Entries in the index are cross-referenced, so any student who has identified a major research theme should be able to find entries under the topic subheadings as well as in the author and title listings.

Notes

1. Some government publications are statistics sources and are listed in the following section of this manual. A complete list of government publications available to depository libraries is available in a looseleaf pamphlet. A recent issue was:

United States. Government Printing Office. Division of Public Documents. List of classes of United States government publications available for selection by depository libraries. Washington, Government Printing Office, October 1, 1968. (E45)

2. Two of the most important references to United States government publications are Boyd and Andriot. There is a great deal of difference between the scope and coverage of Boyd's guide and Andriot's bibliography, as between any guide and bibliography. Boyd is considered a guide to government publications, which means that it shows where to find and how to use them. Andriot is a bibliography, which means it consists of lists of the current serials and other publications of the United States government.

Boyd's reference work is valuable for research because it is a well-organized guide to the history of government printing and the distribution of government publications. It lists the agencies of government, offers some description of their purpose and organization, and lists some of their important publications.

Andriot, on the other hand, is primarily a listing of

the publications of the various government departments, including ephemeral material such as newsletters.   The Andriot bibliography, completed in 1964, lists current government serials produced in the Washington area in the first volume; current notes, news releases and like publications in the second volume; and the publications of agencies outside Washington in volume three.   These three volumes of the 1964 edition and the 1965 Supplement were combined in a 1967 consolidation, updated to January, 1967.   This basic two-volume set will be kept up to date by yearly supplements.

Neither of these two references works is intended as a comprehensive listing of all United States publications. Boyd is valuable because it explains the organization of the government and indicates some of the publications of the various agencies.   Andriot is a bibliography, so it does list publications, but it is designed to be current, so it emphasizes recent publications.   These and other comprehensive guides and catalogs are discussed in Schmeckebier and Eastin (B22).

3.   The Monthly Catalog, indispensable to government document librarians, has a long history beginning in January, 1895.   The Monthly Catalog is the only current comprehensive guide to United States government publications.   The depository library system, outlined by Congress in 1813 to replace a haphazard distribution system practiced earlier, provided government publications to every historical society, college and university library in the United States.   Further resolutions, especially those of 1857, 1858 and 1962 enlarged the number of institutions eligible to receive government publications.   As of 1962, there were permitted two depositories for each state's United States senator and two for each congressional district.   The Monthly Catalog indicates by a black dot those publications which are sent to depository libraries.   A list of depository libraries current as of July 1, 1968, is available in Schmeckebier and Eastin (B22).

4.   The Wilcox manual on the use of state publications is primarily a bibliography of guides to the use of these government publications.   One section of the book lists, state by state, the most important works on state publications.

Wilcox also contains articles written by experts in the field of state government bibliography.   For example, there are selections such as "The Importance of State Libraries," "Statistical Reporting in the States," "List of State Government Organization Charts," "Legislative Digests and Indexes," "State Printing Plants and State Printing Laws," and "Senate and House Journals," all with appropriate supporting bibliographies.

The Library of Congress checklist, on the other hand, is basically a library accessions list. Its prefatory note describes the checklist as "A record of state documents issued during the past five years which are currently received by the Library of Congress." Although the checklist may be designed as a complete record of state publications it is possible that some such publications may be reported late to the Library of Congress or may not be reported at all. To be sure of compiling a complete bibliography, one should consult other records of state publications. Most states maintain their own lists of publications.

For supplementary references on older state publications, see:

Coulter, Edith M. "Selected List of References on State Documents," in Wilcox, Jerome K., (ed.), Manual on the use of state publications, Chicago, American Library Association, 1940. pp. 109-113. (E46)

Bowker, R. R. "Poore's catalogue of government publications." Library Journal, 11:4-5, January 1886. (E47)

Clarks, Edith E. "United States public documents and their catalogs." Library Journal, 31:317-318, July 1906. (E48)

Faison, Georgia H. "Manual on the use of state publications." Social Forces, 19:292, December 1940. (E49)

"Government Publications." Library Journal, 75:2144-2146. December 15, 1950. (E50)

"Government publications and their use." Booklist, 57:451, March 15, 1961. (E51)

Hardin, Ruth. "United States government publications." College and Research Libraries, 12:161, April 1951. (E52)

Hollingsworth, Josephine B. "State documents." Library Journal, 65:696, September 1, 1950. (E53)

"Manual of government publications, United States and

foreign." Booklist, 47:4, September 1, 1950. (E54)

Rosenthal, Joseph A. "Guide to government publications." Library Journal, 86:1868, May 15, 1961.
(E55)

Wyer, J. K., Jr. "Checklist of United States public
documents 1789-1899." Library Journal, 37:630-631,
November 1912. (E56)

Chapter V

STATISTICS SOURCES

Guides to statistics are considered secondary reference works here because they are concerned with only one technique or information source. An important and growing body of social scientists is concerned with the gathering of data and the quantification of problems, so guides to statistics have become extremely important. Statistics are powerful tools in the hands of geographers, and the laws enabling geographers to explain distributions and make spatial predictions are very useful.

Statistics are reported in some bibliographies and compendia which are international in character as well as in more numerous guides to the statistics of individual countries and particular agencies. Many statistical publications are concerned with only one aspect of data collection, such as population, commerce, finance or industry.

General Statistics on the World
and/or its Major Subdivisions

Among the few bibliographies containing general statistical information on the entire world are the following:

United Nations. Statistical office. Statistical papers: Series M. New York, 1949- . No. 1- . Irregular. (F1)

United States. Bureau of the Census. Foreign statistical publications accessions list, January, 1956- . Washington, 1956- . Quarterly. (F2)

The compendia of international general statistics include the following:

Institut International de Statistique. Revue, 1. année 1933- . La Haye, 1933- . (F3)

_____. Office Permanent. Annuaire international de
statistique. La Haye, 1916-1921. 8 vols. (F4)

Statesman's year-book; statistical and historical annual
of the states of the world, 1864- . London and New
York, Macmillan, 1864- . Annual. (F5)

United Nations. Statistical Office. Statistical yearbook;
Annuaire statistique, 1948- . New York, United Nations, 1949- . Annual. (F6)

_____. Monthly bulletin of statistics, no. 1, January
1947- . New York, 1947- . Monthly. (F7)

Among the several compendia, handbooks and bibliographies of statistics pertaining to major subdivisions of the
world are:

United States. Library of Congress. Reference Department. Statistical bulletins; an annotated bibliography
of the general statistical bulletins of major political
subdivisions of the world. Washington, Government
Printing Office, 1954. (F8)

_____. _____. _____. Statistical yearbooks; an annotated
bibliography of the general statistical yearbooks of
major political subdivisions of the world. Washington,
Government Printing Office, 1953. (F9)

Mueller, Bernard. A statistical handbook of the North
Atlantic area. New York, Twentieth Century Fund,
1965. (F10)

Lander der Erde; politisch-ökonomisches Handbuch. Berlin, Verlag Die Wirtschaft, 1962. (F11)

United States. Library of Congress. Census Library
Project. National censuses and vital statistics in
Europe, 1918-1939; an annotated bibliography. Wash-
ington, Government Printing Office, 1948. (F12)

_____. _____. _____. National censuses and vital sta-
tistics in Europe, 1940-1948 Supplement. Washington,
Government Printing Office, 1948. (F13)

Inter-American Statistical Institute. Bibliography of selected statistical sources of the American nations.

Washington, Inter-American Statistical Institute, 1947. (F14)

West Indies and Caribbean yearbook, 1926/27- . London, Skinner, 1927- . Annual. (F15)

California. University. University at Los Angeles. Committee on Latin American Studies. Statistical abstract of Latin America, 1955- . Los Angeles, UCLA, 1955- . Annual. (F16)

America en cifras, 1960- . Washington, Pan American Union, 1961- . Irregular. (F17)

United States. Library of Congress. Census Library Project. General censuses and vital statistics in the Americas. Washington, Government Printing Office, 1943. (F18)

West Africa annual, 1962- . London, J. Clarke, 1962- . Annual. (F19)

Yearbook and guide to East Africa, 1950- . London, R. Hale, 1950- . (F20)

Arab States Fundamental Education Centre. Social Science Division. Statistical sources of the Arab states; a comprehensive list. Sirs-el-layyan, Arab States Training Centre for Education for Community Development, 1961. (F21)

Statistical handbook of Middle Eastern countries. 2nd ed. Jerusalem, Jewish Agency for Palestine, 1945. (F22)

The various agencies of the League of Nations and the United Nations have produced a number of specialized reference works on international statistics. Following are representative specialized reference works produced by the League of Nations and the United Nations.

League of Nations. Statistical yearbook of the League of Nations: 1926-1942/44. Geneva, League of Nations, 1927-1945. Annual. Deceased. (F23)

United Nations. Food and Agriculture Organization. Trade yearbook; Annuaire du commerce; Anuario de

comercio, 1958- . Rome, United Nations, 1959- .
(F24)

_____. Statistical Office. Demographic yearbook. New
York, United Nations, 1950- . Annual. (F25)

_____. _____. World energy supplies. New York,
United Nations, 1951- . Annual. (F26)

_____. Department of Economic and Social Affairs.
World economic survey. New York, United Nations,
1945-47- . Annual. (F27)

_____. Economic and Social Council. Economic Com-
mission for Europe. Economic survey of Europe.
Geneva, United Nations, 1947- . Annual. (F28)

_____. _____. Economic Commission for Latin Amer-
ica. Economic survey of Latin America. New York,
United Nations, 1948- . Annual. (F29)

_____. _____. Economic Commission for Africa. Eco-
nomic bulletin for Africa. New York, United Nations,
1962- . Annual. (F30)

_____. _____. Economic Commission for Asia and the
Far East. New York, United Nations, 1947- . Annual. (F31)

General Statistics on the United States and/or its Major Subdivisions

There are several guides and bibliographies which
lead researchers to general statistics on the United States.
Among these are:

Andriot, John L. Guide to United States Government
statistics. 3rd ed. rev. Arlington, Virginia, Documents Index, 1961. (B28)

United States. Bureau of the Budget. Office of Statistical Standards. Statistical services of the United
States government. rev. ed. Washington, Government Printing Office, 1963. (F32)

_____. Library of Congress. Census Library Project.

Catalog of United States census publications, 1790-
1945. Washington, Government Printing Office, 1950.
(F33)

_____. Bureau of the Census. Bureau of the Census
catalog, 1946- . Washington, Government Printing
Office, 1947- . (F34)

Useful compendia of United States government statistics are:

_____. _____. Statistical abstract of the United States,
1878- . Washington, Government Printing Office,
1879- . vol. 1- . Annual. (F35)

_____. National Vital Statistics Division. Vital statis-
tics of the United States, 1937- . Washington, Government Printing Office, 1939- . Annual. (F36)

_____. _____. County and city data book, 1949- . Wash-
ington, Government Printing Office, 1952- . Irregular.
(F37)

_____. _____. Historical statistics of the United States,
colonial times to 1957. Washington, Government
Printing Office, 1960. (F38)

_____. _____. Historical statistics of the United States,
colonial times to 1957; continuation to 1962 and revisions. Washington, Government Printing Office,
1965. (F39)

The following publications are representative of the
statistical sources available from the cabinet-level offices of
the United States government. Note that they are arranged
alphabetically by cabinet office and, later, by major independent agency. These works are published by the Government Printing Office unless otherwise stated.

1. Department of Agriculture

United States. Department of Agriculture. Agricultural
Statistics. Washington, 1936- . Annual. Succeeds
Annual report 1863-1894, and Yearbooks of Agriculture,
1894-1935. (F40)

_____. _____. Agricultural Conservation Program Ser-

vice. Statistical summary. Washington, 1937- .
Annual. (F41)

_____. _____. Agricultural Marketing Service. Crops
and markets. Washington, 1924- . Annual. (F42)

_____. _____. Agricultural Research Service. Agricul-
tural finance review. Washington, 1938- . Annual.
(F43)

_____. _____. Commodity Stabilization Service. Con-
servation Reserve Program of the Soil Bank. Statistical summary. Washington, 1956- . Annual. (F44)

_____. _____. Commodity Credit Corporation. Charts.
Washington, 1933- . Annual. (F45)

_____. _____. Farmer Cooperative Service. Statistics
of farmer cooperatives. Washington, 19 - . Annual.
(F46)

_____. _____. Federal Extension Service. Extension
activities and accomplishments. Washington, 1940.
Annual. (F47)

_____. _____. Foreign Agricultural Service. Foreign
agricultural trade of the United States. Washington,
1962. Monthly with annual index. (F48)

_____. _____. Forest Service. Tree Planters Notes
Annual Edition. Washington, 1950. Annual. (F49)

_____. _____. _____. Forest fire statistics. Washing-
ton, 1960/61. Annual. (F50)

_____. _____. Rural Electrification Administration.
Annual statistical report. Washington, 1950. Annual.
(F51)

_____. _____. Soil Conservation Service. Soil conser-
vation districts. Washington, 1935. Irregular.
(F52)

2. Department of Commerce

United States. Department of Commerce. Office of

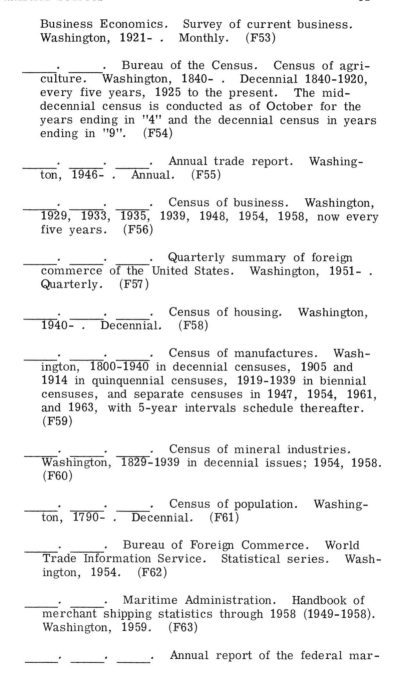

Business Economics. *Survey of current business.*
Washington, 1921- . Monthly. (F53)

_____. _____. Bureau of the Census. *Census of agri-
culture.* Washington, 1840- . Decennial 1840-1920,
every five years, 1925 to the present. The mid-
decennial census is conducted as of October for the
years ending in "4" and the decennial census in years
ending in "9". (F54)

_____. _____. _____. *Annual trade report.* Washing-
ton, 1946- . Annual. (F55)

_____. _____. _____. *Census of business.* Washington,
1929, 1933, 1935, 1939, 1948, 1954, 1958, now every
five years. (F56)

_____. _____. _____. *Quarterly summary of foreign
commerce of the United States.* Washington, 1951- .
Quarterly. (F57)

_____. _____. _____. *Census of housing.* Washington,
1940- . Decennial. (F58)

_____. _____. _____. *Census of manufactures.* Wash-
ington, 1800-1940 in decennial censuses, 1905 and
1914 in quinquennial censuses, 1919-1939 in biennial
censuses, and separate censuses in 1947, 1954, 1961,
and 1963, with 5-year intervals schedule thereafter.
(F59)

_____. _____. _____. *Census of mineral industries.*
Washington, 1829-1939 in decennial issues; 1954, 1958.
(F60)

_____. _____. _____. *Census of population.* Washing-
ton, 1790- . Decennial. (F61)

_____. _____. Bureau of Foreign Commerce. *World
Trade Information Service. Statistical series.* Washington, 1954. (F62)

_____. _____. Maritime Administration. *Handbook of
merchant shipping statistics through 1958 (1949-1958).*
Washington, 1959. (F63)

_____. _____. _____. *Annual report of the federal mar-*

62   Library Research in Geography

itime board and maritime administration. Washington, 1950- . Annual. (F64)

_____. _____. Bureau of Public Roads. Highway statis-
tics: summary to 1955. Washington, 1957. (F65)

_____. _____. _____. Highway statistics. Washington,
1945- . Annual. (F66)

_____. _____. Weather Bureau (ESSA). Climatological
data, national summary. Washington, 1950- . Annual. (F67)

3. Department of Defense

United States. Department of Defense. Annual report.
Washington, 1948- . Annual. (F68)

_____. _____. Corps of Engineers. Waterborne com-
merce of the United States. Washington, 1953- .
Five parts per year. (F69)

_____. _____. Office of the Quartermaster General.
Statistical yearbook of the Quartermaster Corps.
Washington, 1955- . Annual. (F70)

4. Department of Health, Education and Welfare

United States. Department of Health, Education and Welfare. Health, education and welfare trends. Washington, 1960- . Annual. (F71)

_____. _____. Office of Education. Biennial survey of
education in the United States. Washington, 1871- .
Biennial. (F72)

_____. _____. _____. Statistics of land-grant colleges
and universities. Washington, 1869-1870. Annual.
(F73)

_____. _____. Public Health Service. Health statistics
from the United States National Health Survey. Washington, 1958- . Irregular. (F74)

_____. _____. National Office of Vital Statistics. Annual summary of vital statistics. Washington, 1937. Annual. (F75)

_____. _____. Social Security Administration. Annual statistical supplement, social security bulletin. Washington, 1955. Annual. (Prior to 1955, data appeared in September issues.) (F76)

_____. _____. Children's Bureau. Statistical series. Washington, 1947- . (F77)

_____. _____. Bureau of Old-Age and Survivors Insurance. Handbook of old age and survivors insurance statistics. Washington, 1939- . Annual. (F78)

_____. _____. Bureau of Public Assistance. Public assistance, annual statistical data for 1966- . Annual. (F79)

_____. _____. Office of Vocational Rehabilitation. Annual caseload statistics of state rehabilitation agencies. Washington, 1939- . Annual. (F80)

5. Department of the Interior

United States. Department of the Interior. Fish and Wildlife service. Fishery Statistics of the United States. Washington, 1939- . Annual. (F81)

_____. _____. Geological Survey. Water supply papers. Washington, 1896- . Annual and irregular. (F82)

_____. _____. Bureau of Land Management. Public Land Statistics. Washington, 1962- . Annual. Statistical annex of the report of the director. (F83)

_____. _____. Bureau of Mines. Minerals yearbook. Washington, 1932- . Annual. (F84)

_____. _____. National Park Service. State Park service. Washington, 1942- . Annual. (F85)

_____. _____. Bureau of Reclamation. Crop report and related data. Washington, 1963- . Annual. (F86)

6. Department of Justice

United States. Department of Justice. Federal Bureau
of Investigation. Uniform crime reports of the United
States. Washington, 1958. Annual; Quarterly 1930-
40; Semiannually 1941-57. (F87)

_____. _____. Immigration and Naturalization Service.
Annual report. Washington, 1933- . Annual. (F88)

_____. _____. Bureau of Prisons. Federal prisons,
statistical tables. Washington, 1930- . Annual.
(F89)

7. Department of Labor

United States. Department of Labor. Bureau of Employee's Compensation. Federal work injuries sustained during calendar year. Washington, 1963- .
Annual. (F90)

_____. _____. Bureau of Labor Statistics. Monthly
labor review. Washington, 1915- . Monthly. (F91)

_____. _____. Consumer price index. Washington,
1953- . Monthly and quarterly, with annual cumulation. (F92)

8. Post Office Department

United States. Post Office Department. Annual report.
Washington, 1823- . Annual. (F93)

9. Department of State

United States. Department of State. Operations report,
Agency for International Development. Washington,
1962- . Quarterly. (F94)

10. Department of Treasury

United States. Department of Treasury. Annual report
of the secretary of the treasury on the state of the

finances. Washington, 1801- . Annual. (F95)

_____. _____. Office of the Comptroller of the Currency. Annual report. Washington, 1963- . Annual. (F96)

_____. _____. Bureau of Accounts. Combined statement of receipts, expenditures and balances of the United States Government. Washington, 1872- . Annual. (F97)

_____. _____. Bureau of Customs. Merchant marine statistics. Washington, 1924- . Annual. (F98)

_____. _____. Internal Revenue Service. Statistics of Income. Washington, 1916- . Annual. (F99)

_____. _____. Bureau of the Mint. Annual report. Washington, 1873- . Annual. (F100)

11. Major Independent Government Agencies

United States. Civil Aeronautics Board. Annual report, Washington, 1940/1941- . Annual. (F101)

_____. Farm Credit Administration. Annual report. Washington, 1933- . Annual. (F102)

_____. Federal Aviation Agency. Statistical handbook of civil aviation. Washington, 1944- . Annual. (F103)

_____. Federal Communications Commission. Statistics of the communications industry in the United States. Washington, 1939- . Annual. (F104)

_____. Federal Power Commission. Electric power statistics. Washington, 1961- . Monthly. (F105)

_____. Federal Reserve System. Banking and monetary statistics. Washington, 1943- . Annual. (F106)

_____. Interstate Commerce Commission. Annual report. Washington, 1887- . Annual. (F107)

_____. _____. Transport statistics in the United States. Washington, 1954- . Annual. (F108)

_____. Selective Service System. Annual Report. Washington, 1950/1951- . Annual. (F109)

_____. Veterans Administration. Statistical summary of VA activities. Washington, 1931- . Monthly. (F110)

The preceding publications of the United States government agencies are a selection of the statistical works of the cabinet-level departments only. Not included here are the statistical publications of the Executive Office of the President, the Judical branch or the Legislative branch of the United States Government.

Among the bibliographies and guides to state and local statistical publications in the United States are:

United States. Library of Congress. State censuses; an annotated bibliography of censuses of population taken after the year 1790 by states and territories of the United States. Washington, Government Printing Office, 1948. (F111)

_____. Bureau of the Census. State finances, 1942- . Washington, Government Printing Office, 1944- . Annual. (F112)

_____. _____. City finances, 1942- . Washington, Government Printing Office, 1944- . Annual. (F113)

General Statistics on Private Enterprises

Not all statistics are produced by government agencies on their own operations or directed toward students of government. Many statistics sources are available for the student of private economic enterprise. Some of these sources are:

Bogue, Donald J. and Beale, Calvin L. Economic areas of the United States. New York, Free Press of Glencoe, 1961. (F114)

Cole, Arthur H. Measures of business change; a Baker

Library index. Chicago, Irwin, 1952. (F115)

United States. Bureau of Labor Statistics. Monthly labor review, 1915- . Washington, Bureau of Labor Statistics, 1915- . Monthly. (F116)

_____. _____. Office of Business Economics. Survey of current business. Washington, Government Printing Office, 1921- . Monthly. (F117)

_____. _____. Business statistics. Washington, Government Printing Office, 1932- . Biennial. (F118)

Wasserman, Paul, Georgi, Charlotte and Allen, Eleanor. Statistics sources: a subject guide to data on industrial, business, social, educational, financial and other topics for the United States and selected foreign countries. 2nd ed. Detroit, Gale Research Co., 1966. (F119)

Use of Statistics Sources

The foregoing section on statistics has presented two types of statistics sources: 1) bibliographies and guides to statistics, and 2) publications which contain the statistics themselves. The researcher should consult the bibliographies and guides first if he wishes guidance on the types of statistical compilations available on his research topic. If his research topic is well conceived and structured, he may save time in bibliography compilation by proceeding directly to the statistical publications themselves.

Notes

1. Some mention should be made here of the popular one-volume compendia of information, even though they cannot be considered bibliographies or guides. Among the most useful of these "almanacs" are:

World almanac, and book of facts, 1869- . New York, World Telegram, 1868- . vol. 1- . Annual. (F120)

Information please almanac, atlas and yearbook, 1947- . New York, Simon and Schuster, 1947- . Annual. (F121)

Reader's digest almanac, 1966- . Pleasantville, New
York, Reader's Digest Assn., 1966- . Annual. (F122)

2. In many cases guides to national archives are useful in
uncovering specialized statistics. Among the more important
guides to United States government archives are:

United States. National Archives. Guide to the records
in the National Archives. Washington, Government
Printing Office, 1948. (F123)

_____. _____. National Archives accessions, 1947- .
Washington, Government Printing Office, 1947- .
Irregular. (F124)

_____. _____. List of National Archives microfilm pub-
lications, 1965. Washington, Government Printing
Office, 1965. (F125)

_____. _____. Publications of the National Archives and
Records Service. Washington, Government Printing
Office, 1964. (F126)

_____. _____. Your government's records in the Nation-
al Archives. Washington, Government Printing Office,
1950. (F127)

Chapter VI

REFERENCES TO THESES AND DISSERTATIONS

Guides to theses and dissertations are one type of those secondary references which supply information on past and current original research. These guides are among the most important in the reference library because they indicate the trends of research in a discipline as well as supply particular information on the author, title, institution and sometimes content of the work and location where it can be consulted. [1]

Most theses and dissertations produced in the United States are eventually entered in a select few publications. These publications then act as guides to this original research. At times it is necessary to consult abstracts and indexes produced by individual universities to supplement the more comprehensive bibliographies listed below. [2]

United States. Library of Congress. Catalog Division. List of American doctoral dissertations printed in 1912-1938. Washington, Government Printing Office, 1913-1940. 26 vols. (G1)

Doctoral dissertations accepted by American universities, 1933/34-1954/55. Compiled for the Association of Research Libraries. New York, H. W. Wilson, 1934-1956. Nos. 1-22. Deceased. (G2)

Index to American doctoral dissertations, 1955/56- . Compiled for the Association of Research Libraries. Ann Arbor, Michigan, University Microfilms, 1957- . Annual. (G3)

Dissertation abstracts international; abstracts of dissertations and monographs in microform. Ann Arbor, Michigan, University Microfilms, 1952- . vol. 12- . Monthly. (B31)

Dissertations of the sixties. Ann Arbor, Michigan,
University Microfilms, 1964- . Annual. (G4)

Masters abstract: abstracts of selected masters theses
on microfilm. Ann Arbor, Michigan, University Microfilms, 1962- . vol. 1- . Quarterly. (G5)

Black, Dorothy M. (comp.) Guide to lists of master's
theses. Chicago, American Library Association, 1965.
(G6)

                              Notes

1. A number of United States universities cooperate with
University Microfilms, Inc. (Ann Arbor, Michigan) to provide photocopies of completed dissertations.

2. Among other sources of unpublished works of potential
value to English-speaking scholars are:

    Staveley, Ronald. Guide to unpublished research ma-
        terials. London, Library Association, 1957. (G7)

    Palfrey, Thomas R. and Coleman, Henry E. Guide to
        bibliographies of theses, United States and Canada.
        2nd ed. Chicago, American Library Association,
        1940. (G8)

    Marshall, Mary J. Union list of higher degree theses
        in Australian university libraries. Hobart, University
        of Tasmania Library, 1959. (G9)

    Ottawa. Canadian Bibliographic Centre. Canadian grad-
        uate theses in the humanities and social sciences,
        1921-1946. Ottawa, Printer to the King, 1951. (G10)

    Ottawa. National Library of Canada. Canadian theses;
        Theses Canadiennes, 1960/61- . Ottawa, 1962- .
        Annual. (G11)

    Index to theses accepted for higher degrees in the univer-
        sities of Great Britain and Ireland, Vol. 1, 1950/51- .
        London, ASLIB, 1953- . Annual. (G12)

    Recent additions to this list can be gleaned from the
"Dissertations" section of the Bibliographic Index. (A13)

Chapter VII

SUBJECT BIBLIOGRAPHIES IN GEOGRAPHY

Students wishing to compile subject bibliographies in geography are often confused by the wide range of bibliographies, guides, indexes and directories which must be consulted in the course of research. This section of the manual introduces some basic source materials useful to students of geography whatever their current research interest.

These source materials will be presented in the same sequence as that used in previous sections of this manual. First, comprehensive guides to the field, including library catalogs and guides to serials, are presented. Next appear secondary references concerning government publications, statistics sources and theses and dissertations in geography. The final portion of this section contains references to specialized topics in geography.[1]

Comprehensive Guides to Geography

Aside from one standard reference work in geography, little has been done in the United States to produce guides to the literature of the field as a whole.[2] This standard work, identified in the following citation, has been out of print for years.

Wright, John K. and Platt, Elizabeth T. Aids to geographical research: bibliographies, periodicals, atlases, gazetteers and other reference books. 2nd ed. New York, Columbia University Press for the American Geographical Society, 1947. American Geographical Society Research series no. 22. (B1)

The Association of American Geographers recently published the following volume which is not designed specifically for bibliographic purposes, but which may help beginning students find pertinent literature.

Lewthwaite, Gordon R., Price, Edward T., Winters,
Harold A. A geographical bibliography for American
college libraries; a revision of a basic geographical
library, a selected and annotated book list for American colleges. Washington, Association of American
Geographers, 1970. (Commission on College Geography publication no. 9.) (B33)

Printed Library Catalogs

In the United States, the American Geographical Society has been very active in collecting and maintaining a
strong library collection of geography materials.[3] The
American Geographical Society presently has one of the larg-
est libraries of this nature in the world, and the catalog of
its holdings is extremely valuable as a guide to books and
serials in geography. The catalog, listed below, contains a
listing of American Geographical Society library holdings
from its establishment until about 1960.

American Geographical Society of New York. Research
catalogue of the American Geographical Society. Boston,
G. K. Hall, 1962. (B2)

Since 1938, the American Geographical Society has
published a current list of new accessions to its library.
This current list is especially valuable for the researcher
interested in works published after 1960, the year the Research catalogue was completed. This publication is:

American Geographical Society of New York. Current
geographical publications; additions to the Research
catalogue of the American Geographical Society. New
York, American Geographical Society, vol. 1- .
1938- . 10 issues annually. (B3)

Guides to Serials

There are a number of comprehensive lists of the
serials containing the literature of geography.[4] Most important of these in the United States are the following:

Harris, Chauncy D. An annotated world list of selected
current geographical serials in English: including an
appendix of major serials in other languages with reg-

ular supplementary or partial basic use of English.
2nd ed. rev. and enl. Chicago, University of Chicago, 1964. (University of Chicago, Department of
Geography, Research paper no. 96.) (B14)

_____ and Fellmann, Jerome D. International list of
geographical serials. Chicago, University of
Chicago, 1960. (University of Chicago, Department
of Geography, Research paper no. 63.) (B15)

References to Government Publications

There are a number of specialized catalogs and guides
to the publications of the United Nations as well as of the
United States Congress, the Executive branch and the Judici-
ary. Two recent valuable reference works on government
publications of interest to geographers are:

Vinge, C. L. and Vinge, A. G. United States government publications for research and teaching in geography and related social and natural sciences. Totowa,
New Jersey, Littlefield, Adams & Co., 1967. (B19)

Felland, Nordis. "United Nations publications useful to
geographers." The Professional Geographer, 10:11-13,
July 1958. (B20)

Statistics Sources

Of special importance to geographers are these non-
government statistical sources:

Gunther, Edgar, and Goldstein, Frederick A. Current
sources of marketing information; a bibliography of
primary marketing data. Chicago, American Marketing Association, 1960. (H1)

Hauser, Philip M., and Leonard, William R. (eds.),
Government statistics for business use. 2nd ed. New
York, John Wiley and Co., 1956. (H2)

Inter-American Statistical Institute. Bibliography of se-
lected statistical sources of the American nations.
Washington, Inter-American Statistical Institute, 1947.
(F14)

_____. Monthly list of publications received. Washington Inter-American Statistical Institute, 1948- . Monthly. (H3)

National Industrial Conference Board. Economic almanac for 1940- . New York, National Industrial Conference Board, 1940- . Biennial. (H4)

Texas, University. Population Research Center. International population census bibliography. Austin, University of Texas, Bureau of Business Research, 1965-1966. 5 vols. (H5)

The Geographical Digest. London, George Philip and Son, Ltd., 1963- . Annual. (H6)

Wasserman, Paul A., and Georgi, Charlotte, and Allen, Eleanor. Statistics sources: a subject guide to data on industrial, business, social, educational, financial and other topics for the United States and selected foreign countries. 2nd ed. Detroit, Gale Research Co., 1966. (F119)

Weaver, John C., and Lukermann, Fred K. World resources statistics, a geographic sourcebook. 2nd ed. Minneapolis, Burgess, 1953. (H7)

Woytinsky, W. S. Die welt in zahlen. Berlin, 1955. (H8)

_____. World population and production, trends and outlook. New York, Twentieth Century Fund, 1953. (H9)

Among the sourcebooks on statistics useful to geographers are the following:

Anderson, Marc. A working bibliography of mathematical geography. Ann Arbor, Michigan, University of Michigan, Department of Geography, 1963. (Michigan Inter-University Community of Mathematical Geographers Discussion paper no. 2.) (B24)

Berry, Brian J. L., and Hankins, Thomas D. A bibliographic guide to the economic regions of the United States. Chicago, University of Chicago, 1963. (University of Chicago, Department of Geography Research paper no. 87.) (B58)

Berry, Brian J. L., and Pred, Allan R. Central place
studies: a bibliography of theory and applications.
(Philadelphia, Regional Science Research Institute Bibliography series no. 1.) (B25)

Barnum, H. G., Kasperson, R., and Kiuchi, S. Central place studies ... Supplement, 1965. Philadelphia. Regional Science Research Institute, 1965.
(B26)

Olsson, Gunnar. Distance and human interaction: a
review and bibliography. Philadelphia, Regional Science Research Institute, 1965. (Regional Science Research Institute Bibliography series no. 2.) (B42)

Theses and Dissertations

Special compilations of theses and dissertations by
American geographers are available as detailed supplements
to the guides presented in the preceding general section on
theses and dissertations. See the following sources:

Browning, Clyde E. A bibliography of dissertations in
geography: 1901 to 1969. Chapel Hill, North Carolina, University of North Carolina, Department of
Geography, 1970. (Studies in geography no. 1.) (B30)

Whittlesey, Derwent. "Dissertations in geography accepted by universities in the United States for the de-
gres of Ph.D. as of May, 1935." Annals of the Association of American Geographers, 25:211-237, December 1935. (B29)

Hewes, Leslie. "Dissertations in geography accepted by
universities in the United States for the degree of
Ph.D. to June, 1946." Annals of the Association of
American Geographers, 26:215-247, December 1946.
(H10)

_____. "An analysis of doctoral dissertations in geogra-
phy in Anglo America." Newsletter of the Association
of American Geographers, 36:36, June 1947. (H11)

_____. "Dissertations completed and in progress." In
the Professional Geographer, various volumes as follows:

Vol. II   (1950), No. 1, pp. 8-11; No. 3, pp. 11-15;
           No. 4, pp. 14-17; No. 5, pp. 11-12.
Vol. III   (1951), No. 1, pp. 10-12; No. 3, pp. 12-13;
           No. 5, pp. 34-35.
Vol. IV   (1952), No. 4, pp. 26-28; No. 6, pp. 39-40.
Vol. V   (1953), No. 3, pp. 34-36; No. 6, pp. 50-52.
Vol. VI   (1954), No. 2, pp. 55;   No. 6, pp. 21-23.
Vol. VII   (1955), No. 2, pp. 26;   No. 6, pp. 22-24.
Vol. VIII (1956), No. 2, pp. 42;   No. 6, pp. 26-28.
Vol. IX   (1957), No. 2, pp. 51-52; No. 6, pp. 29-30.
Vol. X   (1958), No. 2, pp. 29-30; No. 6, pp. 30-36.
Vol. XI   (1959), No. 2, pp. 39;   No. 6, pp. 33-35.
Vol. XII   (1960), No. 6, pp. 23-25. (H12)

Since 1960, all theses and dissertations completed or
in progress in geography are listed in the November (No. 6)
issues of the Professional Geographer.

Guides to Geography Specialties

Guides to topical, regional and technical specialties in
geography are listed here in that order. When guides to major subdivisions of these fields are available, they are included. Not all guides listed are written by geographers or
for geographers, but all are helpful in geographic research. [5]

1. Topical geography

a. Physical geography

1. Geomorphology

Pearl, Richard M.   Guide to geologic literature.   New York,
    McGraw-Hill Book Co., 1951.   (H13)

Ward, Dederick C.   Geologic reference sources.   Metuchen,
    N.J.: Scarecrow Press, 1972.   (B34)

International Geographical Union.   Commission on Coastal
    Sedimentation.   Bibliography 1955-1958.   Baton Rouge,
    Louisiana, Louisiana State University Coastal Studies In-
    stitute, 1960.   (H14)

2. Meteorology and Climatology

American Meteorological Society. Meteorological and geoastrophysical abstracts. Boston, American Meteorological
Society, 1950- . Monthly. (B37)

United States. Department of Commerce. Weather Bureau.
Bibliographies of climatic maps. Washington, Government Printing Office, Nos. 1-40, 1958-1962. (H15)

_____. _____. Bibliographies on climate. Washington, Gov-
ernment Printing Office, 1957-1962. (H16)

_____. Weather Bureau. Selective guide to published climatic
data sources. Washington, Government Printing Office,
1969. (Key to meteorological records documentation no.
4.11.) (B35)

3. Hydrology

American Geophysical Union. Bibliography of hydrology,
United States of America. Washington, American Geophysical Union, 1940. (H17)

American Water Resources Association. Hydata, 1965- .
Urbana, Illinois, American Water Resources Association,
1965- . Monthly. (H18)

Water Resources Council. Hydrology and Sedimentation Committees. Annotated bibliography on hydrology and sedimentation, 1963-1965, United States and Canada. Washington, Water Resources Council, 1969. (Joint hydrology-
sedimentation bulletin no. 9.) Supplement to previous
bibliographies on hydrology and sedimentation that were
prepared in cooperation with or under the auspices of the
Subcommittee on Hydrology and Sedimentation, Inter-Agency
Committee on Water Resources. Published through arrangements made by the Soil Conservation Service acting
in behalf of several participating governmental agencies.
(B36)

4. Oceanography

Interagency Committee on Oceanography. Bibliography of
oceanographic publications. ICP Pamphlet No. 9, Washington, Interagency Committee on Oceanography, 1963.
(H19)

National Research Council. Committee on Oceanography.
Oceanographic information sources; a staff report. Wash-
ington, National Academy of Sciences-National Research
Council, 1970. (B38)

Massachusetts Oceanographic Institution (Woods Hole). A
partial bibliography of the Indian Ocean. Woods Hole,
Massachusetts, Woods Hole Oceanographic Institution,
1962. (H20)

5. Biogeography

Blake, Sidney F. Geographical guide to floras of the world.
New York, Hafner, 1963. (Reprint of the 1942 edition of
the United States Department of Agriculture, Miscellaneous
publication No. 401.) (B39)

Jackson, Benjamin D. Guide to the literature of botany.
New York, Hafner Publishing Co., 1964. (H21)

Kerker, Ann E. and Schlundt, Esther M. Literature sources
in the biological sciences. Lafayette, Indiana, Purdue
University, 1961. (H22)

Smith, Roger C. Guide to the literature of the zoological
sciences. 6th ed. Minneapolis, Burgess, 1962. (B40)

6. Soils

Commonwealth Bureau of Soils. Bibliography of soil science,
fertilizers and general agronomy, 1959-1962. Farnham
Royal, England, Commonwealth Agricultural Bureaux,
1964. (B41)

b. Cultural geography

1. Urban

Berry, Brian J. L. and Pred, Allan. Central place studies:
a bibliography of theory and applications. Philadelphia,
(Regional Science Research Institute Bibliography series
no. 1.) (B25)

Barnum, H. G., Kasperson, R., and Kiuchi, S. Central
place studies ... Supplement, 1965. Philadelphia, Regional Science Research Institute, 1965. (B26)

Bestor, George C., and Jones, Holway R.   City planning:
a basic bibliography of sources and trends.   Rev. ed.
Sacramento, California Council of Civil Engineers and
Land Surveyors, 1966.   (B43)

Branch, Melville C.   Comprehensive urban planning: a selective annotated bibliography with related materials.
Beverly Hills, California, Sage, 1970.   (B44)

Chapin, F. Stuart, Jr.   Selected references on urban planning, methods and techniques.   Durham, North Carolina,
University of North Carolina, Department of City and Regional Planning, 1963.   (H23)

Holleb, Doris B.   Social and economic information for urban
planning.   Chicago, Center for Urban Studies of the University of Chicago, 1969. 2 vols.   (B45)

2. Housing

National Association of Home Builders of the United States.
Basic texts and reference books on housing and construction: a selected, annotated bibliography.   Washington, National Association of Home Builders, 1956.   (B46)

National Association of Housing and Redevelopment Officials.
Summary of the housing year; bibliography of housing lit-
erature.   Chicago, National Housing and Redevelopment
Officials, 19 - .   (H24)

3. Population

Eldridge, Hope T.   The materials of demography: a selected and annotated bibliography.   New York, Columbia
University Press, 1959.   (H25)

Princeton University.   Office of Population Research and
Population Association of America.   Population Index.
Princeton, 1935- .   Quarterly.   (H26)

Zelinsky, Wilbur.   A bibliographic guide to population geography.   Chicago, University of Chicago, 1962.   (University
of Chicago, Department of Geography, Research paper no.
80.)   (B47)

4. Industrial

National Association of Manufacturers of the United States of

America. Bibliography of economic and social study material issued by the National Association of Manufacturers. New York, National Association of Manufacturers, 1940-1946. (H27)

Stevens, Benjamin H., and Brackett, Carolyn A. Industrial location: a review and annotated bibliography of theoretical, empirical and case studies. Philadelphia, Regional Science Research Institute, 1967. (Regional Science Research Institute Bibliography series no. 3.) (B48)

Wilson, Fern L. Index of publications by University bureaus of business research. Cleveland, Western Reserve University Press, 1951-1957. (H28)

United Nations. Headquarters Library. Bibliography on industrialization in under-developed countries. United Nations library bibliographical series No. 6. New York, United Nations, 1956. (H29)

   5. Agriculture

Blanchard, J. R. and Ostvold, Harold. The literature of agricultural research. Berkeley, University of California Press, 1958. (H30)

Lauche, Rudolf (ed.) World bibliography of agricultural bibliographies. Munich, Bayerischer Lanwirtschaftsverlag, 1957. (H31)

United States. Department of Agriculture. Index to publications of the United States Department of Agriculture. Washington, Government Printing Office, 1901- . (H32)

_____. _____. Bibliography of agriculture. Washington, Government Printing Office, 1961- . Six issues yearly. (H33)

Wang, Jen Yu, and Barger, Gerald L. (eds. and comps.) Bibliography of agricultural meteorology. Madison, University of Wisconsin Press, 1962. (B49)

   6. Transportation

Blaisdell, Ruth F. et. al. Sources of information in transportation. Evanston, Northwestern University Press, 1964. (H34)

Metcalf, Kenneth H.   Transportation information sources; an
   annotated guide to publications, agencies, and other data
   sources concerning air, rail, water, road and pipeline
   transportation.   Detroit, Gale Research Co., 1965.   (H35)

Northwestern University.   Library of the Transportation Cen-
   ter.   Current literature in traffic and transportation.
   Evanston, Northwestern University Transportation Center
   Library, 1960- .   Monthly.   (H36)

Siddall, William R.   Transportation geography: a bibliography.   Manhattan, Kansas, Kansas State University, 1967.
   (Kansas State University Library Bibliography series no.
   1.)   (B50)

United States.   Department of Commerce.   Bureau of Public
   Roads.   Highways: current literature.   Vol. 1- .   1921- .
   Weekly.   (H37)

Wolfe, Roy I. and Hickok, Beverly.   An annotated bibliography of the geography of transportation.   Institute of Transportation and Traffic Engineering Information Circular No.
   29.   Berkeley, University of California Press, 1961.   (H38)

   7.   Land Use and Land Tenure

Bercaw, Louise O.   Bibliography on land utilization 1918-
   1936.   Washington, Government Printing Office, 1938.
   (H39)

Centre for Urban Studies.   Land use planning and the social
   sciences: a selected bibliography.   London, Centre for
   Urban Studies, 1964.   (H40)

Culver, Dorothy C.   Land utilization: a bibliography.
   Berkeley, University of California Bureau of Public Administration, 1937.   (H41)

Denman, Donald R.   Bibliography of rural land economy and
   land ownership, 1900-1957.   Cambridge, Massachusetts
   Harvard University Department of Estate Management,
   1958.   (H42)

Food and Agriculture Organization of the United Nations.
   Bibliography on land tenure.   New York, United Nations,
   1955.   (B51)

Food and Agriculture Organization of the United Nations.
Bibliography on land tenure. New York, United Nations,
1959. Supplement. (H43)

Wisconsin University. Land Tenure Center. Accessions
lists. Madison, University of Wisconsin Land Tenure
Center, 1963- . Irreg. (H44)

8. Economic Development

Hazlewood, Arthur. (comp.) The economics of "underdeveloped" areas: an annotated reading list of books, articles
and official publications. 2nd ed. London, Institute of
Commonwealth Studies, 1959. (H45)

Katz, Saul M. and McGowan, Frank. A selected list of
readings on development. Washington, Agency for International Development, 1963. (H46)

ReQua, Eloise G. and Statham, Jane. The developing nations: a guide to information sources concerning their
economic, political, technical and social problems. Detroit, Gale Research Company, 1965. (Management information guide series no. 5.) (B52)

Stewart, Charles F. and Simmons, George B. (comps.)
A bibliography of international business. New York, Columbia University Press, 1964. (H47)

United Nations. Economic and Social Council. International
bibliography of economics. New York, United Nations,
Vol. 1- . 1955- . Annual. (H48)

9. Political

Burdette, Franklin L. (and others). Political science: a selected bibliography of books in print with annotations. College Park, Maryland, University of Maryland Bureau of
Government Research, 1961. (H49)

Harmon, Robert B. Political science: a bibliographical
guide to the literature. New York, Scarecrow, 1965.
(B53) (Supplements, 1968 and 1972)

United Nations. Economic and Social Council. International
bibliography of political science. New York, United Nations. Vol. 1- . 1954- . (H50)

10. Military

Geological Society of America. Bibliography of military
  geology and geography. New York, Geological Society of
  America, 1941. (H51)

Peltier, Louis C. (ed.) Bibliography of military geography.
  Washington, Association of American Geographers, 1962.
  (B54)

11. Historical

American Bibliographical Center. America: history and
  life, a guide to periodical literature. Santa Barbara,
  California, American Bibliographical Center, Vol. 1- .
  1964- . (H52)

McManis, Douglas R. Historical geography of the United
  States: A bibliography--excluding Alaska and Hawaii.
  Ypsilanti, Michigan, Eastern Michigan University, 1965.
  (B55)

2. Regional Geography

a. General

American Universities Field Staff. A select bibliography:
  Asia, Africa, Eastern Europe, Latin America. New York,
  American Universities Field Staff, 1960 (and supplements).
  (B57)

_____. Supplement, 1961. (H53)

_____. Supplement, 1963. (H54)

Logan, Marguerite. Geographical bibliography for all the
  major nations of the world; selected books and magazine
  articles. Ann Arbor, University of Michigan Press, 1959.
  (H55)

Olson, Ralph E. The literature of regional geography; a
  checklist for university and college libraries. Norman,
  Oklahoma, National Council for Geographic Education,
  1960. (H56)

b. Anglo-America

Staton, Frances M. A bibliography of Canadiana; being
    items in the public library of Toronto, Canada, relating
    to early history and development of Canada. Toronto,
    Toronto Public Library, 1934. (H57)

Toronto. Public Library. The Canadian catalogue of books
    published in Canada, about Canada, as well as those writ-
    ten by Canadians, 1921-1949. Toronto, Toronto Public
    Library, 1959. (H58)

United States. Library of Congress. A guide to the study
    of the United States of America; representative books re-
    flecting the development of American life and thought.
    Washington, Library of Congress, 1960. (H59)

c. Latin America

Florida. University and United States. Library of Congress.
    Hispanic Foundation. Handbook of Latin American Studies.
    Gainesville, University of Florida Press, 1936- . Pub-
    lisher varies. Annual. (H60)

Gropp, Arthur E. (comp.) A bibliography of Latin American
    bibliographies. Washington, Pan American Union, 1968.
    (B59)

Hilton, Ronald. Handbook of hispanic source materials and
    research organizations in the United States. Palo Alto,
    Stanford University Press, 1956. (H61)

d. Europe

Conover, Helen F. (comp.) Introduction to Europe; a se-
    lective guide to background reading. Washington, Library
    of Congress, European Affairs Division, 1950. (H62)

European Cultural Centre. The European bibliography, ed.
    by Hjalmar Pehrsson and Hanna Wulf. Leyden, A. W.
    Sijthoff, 1965. (H62)

e. Soviet Union

Horecky, Paul L. (ed.) Basic Russian publications, an an-
    notated bibliography on Russia and the Soviet Union. Chi-
    cago, University of Chicago Press, 1962. (H63)

Maichel, Karol. Guide to Russian reference books. Hoover
   Institution on War, Revolution and Peace, Bibliography
   series, 10 and 18. Palo Alto, California, Stanford University, 1962 and 1964. (H64)

United States. Library of Congress. Reference Division.
   Soviet geography, a bibliography. Ed. by Nicholas R.
   Rodionoff. Washington, Government Printing Office, 1951.
   2 vols. (B60)

f. Africa

African Bibliographic Center. A current view of Africana;
   a select and annotated bibliographic publishing guide.
   Washington, African Bibliographical Center, 1964. (H65)

Twentieth Century Fund. Survey of tropical Africa; select
   annotated bibliography of tropical Africa. ed. by C.
   Daryll Forde. New York, International African Institute,
   1956. (B61)

United States. Library of Congress. African Section.
   Africa south of the Sahara: a selected, annotated list of
   writings. Comp. by Helen R. Conover. Washington, Government Printing Office, 1963. (H66)

g. Asia

Embree, John F. Books on Southeast Asia: a select bibliography. New York, American Institute of Pacific Relations, 1956. (H67)

Field, Henry H. Bibliography on Southwestern Asia. Coral
   Gables, Florida, University of Miami Press, 1953. (H68)

Hall, Robert B., and Noh, Toshio. Japanese geography: a
   guide to Japanese reference and research materials. Ann
   Arbor, Michigan, University of Michigan Press, 1956.
   (University of Michigan Center for Japanese Studies Bibliographic series no. 6.) (B62)

Pelzer, Karl J. Selected bibliography on the geography of
   Southeast Asia. New Haven, Yale University, 1949-1956.
   3 vols. (B63)

h. Australia - New Zealand - Pacific

Australia. National Library. Australian national bibliogra-

phy. Canberra, National Library of Australia, 1961.
Monthly. (H69)

Harris, William J. Guide to New Zealand reference material and other sources of information. 2nd. ed. Wellington, New Zealand Library Association, 1950. (H70)

Sachet, Marie H. and Fosberg, Francis R. Island bibliographies: Micronesian botany, land environment and ecology of coral atolls, vegetation of tropical Pacific Islands.
Washington, National Academy of Sciences--National Research Council, 1955. (H71)

Taylor, Clyde R. H. A Pacific bibliography: printed material relating to the native peoples of Polynesia, Melanesia and Micronesia. Wellington, New Zealand, Polynesian
Society, 1951. (H72)

i. Polar

The Arctic Institute of North America. Arctic bibliography,
Vol. 1- . 1952- . (H73)

United States. National Science Foundation. Office of Antarctic Programs. Antarctic bibliography. Washington,
Government Printing Office, 1966. (H74)

j. Tropics

International Geographical Union. Special Commission on the
Humid Tropics. A select annotated bibliography of the
humid tropics. Comp. by Theo L. Hills. Montreal,
McGill University, 1960. (B56)

Guides to Maps and Atlases

Although no one knows how many maps and atlases
are produced in the world, there are several bibliographic
sources which list the cartographic production of the United
States.[6] Included here are bibliographies, indexes, library
catalogs, accessions lists and other important sources.

American Geographical Society. Map Department. Index
to maps in books and periodicals. Boston, G. K.
Hall, 1967. (B65)

British Museum. Department of Printed Books. Map
Room. Catalogue of printed maps, plans and charts
in the British Museum. London, British Museum,
1885. 2 vols. (H75)

California. University. Berkeley. Bancroft Library.
Index to printed maps. Boston, G. K. Hall, 1964.
(H76)

Claussen, Martin P. and Friis, Herman R. Descriptive
catalog of maps published by Congress, 1817-1843.
Washington, Printed privately, 1941. (H77)

Porter, Philip W. A bibliography of statistical cartography. Minneapolis, University of Minnesota Book-
store, 1964. (B64)

Thiele, Walter. Official map publications: a historical
sketch, and a bibliographical handbook of current
maps and mapping services in the United States,
Canada, Latin America, France, Great Britain, Germany and certain other countries. Chicago, American
Library Association, 1938. (H78)

United Nations. Food and Agriculture Organization.
Catalogue of maps. 3rd ed. Rome, United Nations,
1965. (H79)

United States. Library of Congress. Library of Congress catalogue: a cumulative list of works represented by Library of Congress printed cards: Maps
and Atlases. Washington, Library of Congress,
1953-1955. 3 vols. Deceased. (H80)

_____. _____. Map Division. A guide to historical
cartography, a selected, annotated list of references on the history of maps and map making,
comp. by Walter W. Ristow and Clara E. Le-
Gear. 2nd ed. rev. Washington, Library of Congress, 1960. (H81)

_____. _____. _____. A list of geographical atlases in
the Library of Congress, with bibliographical notes.
Washington, Government Printing Office, 1909-1963.
6 vols. In progress. (H82)

_____. _____. _____. United States Atlases: a list of

national, state, county, city and regional atlases in
the Library of Congress, comp. by Clara E. LeGear.
Washington, Government Printing Office, 1950-1953.
2 vols. (H83)

_____. _____. _____. Aviation cartography: a historico-
bibliographic study of aeronautical charts, by Walter
W. Ristow. 2nd ed. rev. enl. Washington, Government Printing Office, 1960. (H84)

Information on current international and United States
map and atlas production can be gleaned from the following
several publications.[7] Both international and United States
publications are listed in this section.

American Geographical Society. Library. Current geographical publications: additions to the Research catalogue of the American Geographical Society. New
York, American Geographical Society, Vol. 1- .
1938- . 10 issues yearly. (B3)

Bibliographie cartographique internationale, 1936- .
Paris, Armand Colin, 1938- . Annual. Irregular.
(H85)

Edward Stanford, Ltd. International map bulletin. London, Edward Stanford, 1947- . Annual. (H86)

German Cartographical Society and the Bundesanstalt fur
Landeskunde und Raumforschung. Bibliotheca car-
tographica. Bad Godesbarg, Germany, Rundesanstalt
fur Landeskunde und Raumforschung, 1957- . Semiannual. (H87)

International Hydrographic Bureau. International Hydrographic Bulletin. Supplement to the International Hydrographic Review. Monaco, International Hydrographic Bureau, 1928- . Monthly. See "List of New
Charts and New Editions of Charts" Section. (H88)

Journal of Geography. Chicago, National Council for
Geographic Education, 1902- . Monthly except June,
July and August. (H89)

Military Engineer. Washington, Society of American
Military Engineers, 1920- . 6 issues per year.
(H90)

Professional Geographer. Washington, Association of
American Geographers, 1949- . 6 issues yearly.
(H91)

Rand McNally and Company. Geographic Research Department. Map library acquisitions bulletin: a list
of atlases, maps and books received by the map library. Chicago, Rand McNally, 19 - . Irregular.
(H92)

Royal Geographical Society. New Geographical literature
and maps: additions to the library of the Royal Geographical Society, 1951- . London, Royal Geographical Society, 1951- . Semiannual. (H93)

Special Libraries Association. Geography and Map Division Bulletin. Washington, Special Libraries Association, 1950- . Quarterly. (H94)

Surveying and Mapping. Washington, American Congress
on Surveying and Mapping, 1941- . Quarterly. (H95)

United Nations. World cartography. New York, United
Nations, 1950- . Annual. (H96)

United States. Library of Congress. Copyright Office.
Catalog of copyright entries: third series; part 6:
Maps and atlases. Washington, Government Printing
Office, 1947- . Semiannual. (H97)

_____. _____. Processing Department. Monthly check-
list of state publications, 1910- . Washington, Government Printing Office, 1910- . Monthly. (E36)

_____. Navy. Oceanographic Office. DoD Nautical
Chart Library. Accession list of domestic and foreign
charts, cumulative list. Washington, Naval Oceanographic Office, 19 - . Monthly. (H98)

_____. Superintendent of Documents. Monthly catalog
of United States government publications, 1895- .
Washington, Government Printing Office, 1895- .
Monthly. See entries for federal mapping agencies.
(E21)

Some important reviews of and guides to maps and
atlases are the following:

Hannah, William. "Foreign topographic mapping agencies and their sales and information offices." Surveying and Mapping, 16:506-508, October-December 1956, 17: 200-201, April-June 1957. (H99)

Nicholson, N. L. "A survey of single-country atlases." Geographical Bulletin, 2:19-35, 1952. (H100)

Ristow, Walter W. "The emergence of maps in libraries." Special Libraries, 58:400-419, July-August, 1967. (H101)

Stephenson, Richard W. "Published sources of information about maps and atlases." Special Libraries, 61: 87-112, February 1970. (H102)

Yonge, Ena L. "National atlases: a summary." Geographical Review, 52:407-432, 1962. (H103)

_____. "Regional atlases: a summary survey." Geographical Review, 52:407-432, 1962. (H104)

_____. "World and thematic atlases: a summary survey." Geographical Review, 52:583-596, 1962. (H105)

Guides to Photogrammetry, Photo Interpretation
and Remote Sensing

American Society of Photogrammetry. Manual of photographic interpretation. Washington, American Society of Photogrammetry, 1960. (H106)

Fischer, William A. Remote sensing research in the United States: a summary. Proceedings XI International Congress for Photogrammetry, Lausanne, Switzerland, July 1968. (H107)

Gywer, Joseph A. and Waldron, Vincent G. Photo interpretation techniques; a bibliography. Washington, Library of Congress 1956. (B66)

NASA. Earth resources research data facility index. Houston, Manned Spacecraft Center, July 1969. (H108)

_____. Thermal infrared imagery in urban studies. Technical letter NASA-135, Houston: Manned Spacecraft Center, December, 1968. (H109)

_____. Uses of conventional aerial photography in urban areas: review and bibliography. Technical letter NASA-131, Houston, Manned Spacecraft Center, September 1968. (H110)

Walters, Robert L. Radar bibliography for geoscientists. Lawrence, Kansas, University of Kansas, 1968. (CRES Report no. 61-30, Remote Sensing Laboratory.) (B67)

Guides to Geography Education

Anderzhon, Mamie L. A selected bibliography on geography education for curriculum committees. Norman, Oklahoma, National Council for Geographic Education, 1963. (National Council for Geographic Education Special Publication No. 7.) (H111)

Gabler, Robert E. (ed.) A handbook for geography teachers. Normal, Illinois, National Council for Geographic Education, 1966. (Geographic education series no. 6.) (B68)

Proctor, Cleo V., Jr. "Paperback books for earth science teachers." Journal of Geological Education, 15: 29-55, 1967. (H112)

Notes

1. Geology has been allied with geography for many years, and geographers may find references to geology publications useful in bibliographic search. Among the most valuable guides to the literature of geology are the following:

Darton, N. H. "Catalogue and index of contributions to North American geology." United States Geological Survey Bull. 127, 1896. Covers the years 1732-1891. (H113)

Nickles, J. M. "Geologic literature of North America." United States Geological Survey Bull. 746, 1923; Bull. 747, 1924. Covers the years 1785-1918. (H114)

_____. "Bibliography of North American geology."

United States Geological Survey Bull. 823, 1931. Covers the years 1919-1928. (H115)

Thom, E. M. "Bibliography of North American geology." United States Geological Survey Bull. 937, 1965. Covers the years 1929-1939. 2 vols. (H116)

King, R. R., and others. "Bibliography of North American geology." United States Geological Survey Bull. 1049, 1957. Covers the years 1940-1949. 2 vols. (H117)

United States. Geological Survey. "Bibliography of North American geology." United States Geological Survey Bull. 1195, 1965. Covers the years 1950-1959. 4 vols. (H118)

Clarke, J. S., and others. "Bibliography of North American geology." United States Geological Survey Bull. 1196, 1964. Covers 1960. (H119)

United States. Geological Survey. "Bibliography of North American geology." United States Geological Survey Bull. 1197, 1965. Covers 1961. (H120)

_____. _____. "Bibliography of North American geology." United States Geological Survey Bull. 1232, 1966. Covers 1962. (H121)

_____. _____. "Bibliography of North American geology." United States Geological Survey Bull. 1233, 1968. Covers 1963. (H122)

_____. _____. "Bibliography of North American geology." United States Geological Survey Bull. 1234, 1966. Covers 1964. (H123)

_____. _____. "Bibliography of North America geology." United States Geological Survey Bull. 1235, 1969. Covers 1965. (H124)

_____. _____. Abstracts of North American geology. Washington, United States Geological Survey, January, 1966- . Monthly. (H125)

Geological Society of America. Bibliography and index of geology exclusive of North America. Boulder, Col-

orado, Geological Society of America, 1932-1968.
Annual. (H126)

_____. Bibliography and index of geology. Boulder,
Colorado, Geological Society of America, January,
1969- . Monthly; annual cumulations. (H127)

Annotated bibliography of economic geology. Urbana,
Illinois, Economic Geology Publishing Co., 1928- .
Annual. (H128)

Ward, Dederick C. Geologic reference sources. Boulder, University of Colorado Press, 1967. (University
of Colorado series in earth sciences no. 5.) (B34)

Harvey, Anthony P. "Recent developments in geological
documentation and bibliography." Proceedings, Geoscience Information Society, 1:27-53, 1969. (H129)

Long, H. K. "Bibliography of state geological bibliographies." Geoscience Abstracts, 7:111-125, 1965. (H130)

Chronic, J., and Chronic, H. Bibliography of theses in
geology. Boulder, Colorado, Pruett Press, 1958.
Covers through 1957. (H131)

_____. Bibliography of theses in geology. Washington,
American Geological Institute, 1964. (H132)

Ward, Dederick C. "Bibliography of theses in geology."
Geoscience Abstracts, 7:101-137, 1965. (H133)

_____. Bibliography of theses in geology. Washington,
American Geological Institute, 1969. (H134)

Geoscience Information Society. Geologic field trip guide
books of North America. Houston, Phil Wilson, 1968.
(H135)

American Chemical Society. Chemical abstracts. Washington, American Chemical Society, 1907- . Biweekly.
(H136)

Engineering Index, Inc. Engineering index. New York,
Engineering Index, Inc., 1884. Monthly. (H137)

Zoological Society of London. Zoological record. Lon-

don, Zoological Society of London, 1864- . Monthly.
(H138)

Mineralogical Society of Great Britain and Mineralogical
Society of America. Mineralogical abstracts. London, Mineralogical Society of Great Britain and Mineralogical Society of America, 1920- . Issued every
two years. (H139)

Oceanic Research Institute. Oceanic index. La Jolla,
California, Oceanic Research Institute, 1964- . Various names and formats. Sections appear monthly and
quarterly. (H140)

For more recent information on documentation systems for geology see:

Harvey, Anthony P. "Recent developments in geological
documentation and bibliography." Proceedings of the
Geoscience Information Society, 1:27-53, 1969. (H141)

Smith, Harriet W. "Guide to geological literature."
Journal of Geological Education, 18:13-25, 1970. (H142)

2. Other organizations publish similar guides in foreign
countries. The most important of these are produced by the
Soviet Union, Great Britain, France and the two Germanies.
These sources are as follows:

Akademiia Nauk USSR. Institut Nauchnoi Informatsii.
Referativng; zhurnal: geografiia. 1954- . 12 issues
per year. (H143)

Geographical Abstracts. University of East Anglia, England. 1966- . Four series (Geomorphology; Biogeography, Climatology) with six issues per year. (H144)

Association de Géographes Français. Bibliographie géographique internationale. 1891- . Annual. Vols.
1-24 appear as supplements to Annales de Géographie.
(H145)

Société de Géographie de Paris. Acta géographica. Supplement Bibliographique. 1949- . Quarterly. (H146)

Geographisches Jahrbuch. VEB Hermann Haack, Gotha.
Vol. 1-62, 1866-1967. Deceased. (H147)

Westermann's Geographische Bibliographie. Georg Westermann Verlag, Braunschweig, 1954-1965. Deceased. (H148)

Institut fur Landeskunde. Documentatio Geographica. Geographische zeitschriften-und serien-literatur. Institut fur Landeskunde, Bad Godesberg. 1966- . 6 issues per year. (H149)

Institut fur Landeskunde. Abteilung dokumentation. ZWL-Dokumentationsdienst Regionale Geographie. 1957- . 200 cards per month. (H150)

3. Geographical associations in foreign countries also produce catalogs of their libraries. Among these are:

Royal Geographical Society. New geographical literature and maps: additions to the library of the Royal Geographical Society, 1951- . (Preceded by Recent Geographical literature, maps and photographs. Nos. 1-64. 1918-1941.) London: Royal Geographical Society, 1951- . Semiannual. (H151)

Real Sociedad Geografica. Catalogo de la biblioteca. Madrid, Real Sociedad Geographica, 1947-1948. 2 vols. (H152)

4. Among the more useful of the guides to serials published outside the United States are the following:

Centre National de la Recherche Scientifique. Repertoire des principaux périodiques d'intéret géographique cités dans la Bibliographie Géographique Internationale. Paris, Editions du Centre National de la Recherche Scientifique, 1966. (H153)

Royal Geographical Society. Current geographical periodicals: a hand-list and subject index of current periodicals in the library of the Royal geographical society. London, Royal Geographical Society. London, Royal Geographical Society, 1961. (H154)

Institut fur Landeskunde. Verzeichnis der geographischen zeitschriften, periodischen veroffentlichengen und schriftenreihen Deutschlands und der in den litzteren erschienenen arbeiten. Sonderheft 7, 1964. (H155)

5. At times textbooks and other similar sources provide specialized bibliographies. The following reading list is designed to be indicative of these secondary works. This selected list cannot replace personal initiative in reading, but it can serve as a guide and a stimulus for anyone who wishes to enlarge his understanding of geography. The order of the readings is the same as that of the preceding section on subject bibliographies in geography except that a special section of primary readings in the recent history and philosophy of geography is appended. In the latter section some abbreviations are used in referring to the journal articles; the term "Annals," for example, refers to the Annals of the Association of American Geographers.

READING LIST

1. Topical Geography

a. Physical Geography

  1. Geomorphology

Fenneman, Nevin M. Physiography of the Eastern United
  States. New York, McGraw-Hill, 1931. (H156)

_____. Physiography of the Western United States. New
  York, McGraw-Hill, 1931. (H157)

King, Phillip B. The evolution of North America. Princeton, New Jersey, Princeton University Press, 1959.
  (H158)

Lobeck, Armin K. Geomorphology. New York, McGraw-
  Hill, 1939. (H159)

Putnam, William C. Geology. New York, Oxford University
  Press, 1964. (H160)

Thornbury, William D. Principles of geomorphology. New
  York, Wiley, 1954. (H161)

_____. Regional geomorphology of the United States. New
  York, Wiley, 1965. (H162)

  2. Meteorology and Climatology

Critchfield, Howard J. General climatology. Englewood
  Cliffs, New Jersey, Prentice-Hall, 1960. (H163)

Hare, F. Kenneth. The restless atmosphere. Rev. ed.
  New York, Hillary House, 1961. (H164)

Kendrew, Wilfred G. The climates of the continents. 5th
  ed. Oxford, Clarendon Press, 1961. (H165)

Koeppe, Clarence E., and DeLong, George C. Weather and

climate. New York, McGraw-Hill, 1958. (H166)

Landsberg, Helmut E. Physical climatology. 2nd ed. DuBois, Pennsylvania, Gray Printing Co., 1958. (H167)

Petterssen, Sverre. Introduction to meteorology. 2nd ed. New York, McGraw-Hill, 1958. (H168)

Riehl, Herbert. Introduction to the atmosphere. New York, McGraw-Hill, 1965. (H169)

Trewartha, Glenn T. An introduction to climate. 4th ed. McGraw-Hill, 1968. (H170)

3. Hydrology

Katzman, Raphael. Modern hydrology. New York, Harper and Row, 1965. (H171)

Kuenen, Philip H. Realms of water; some aspects of its cycle and nature. New York, Wiley, 1955. (H172)

Meinzer, Oscar E. (ed.) Hydrology. New York, Dover, 1942. (H173)

Todd, David K. Ground water hydrology. New York, Wiley, 1959. (H174)

4. Oceanography

Carson, Rachel L. The sea around us. Rev. ed. New York, Oxford University Press, 1961. (H175)

King, Cuchlaine A. M. An introduction to oceanography. New York, McGraw-Hill, 1963. (H176)

Sverdrup, Harold U., Johnson, Martin W., and Fleming, Richard H. The oceans: their physics, chemistry, and general biology. New York, Prentice-Hall, 1942. (H177)

Von Arx, William S. Introduction to physical oceanography. Reading, Massachusetts, Addison-Wesley, 1962. (H178)

5. Biogeography

Eyre, Samuel R. Vegetation and soils, a world picture. Chicago, Aldine, 1963. (H179)

Dansereau, Pierre M. Biogeography: an ecological perspective. New York, Ronald, 1952. (H180)

Darlington, Philip J. Zoogeography: the geographical distribution of animals. New York, Wiley, 1957. (H181)

Polunin, Nicholas V. Introduction to plant geography and
some related sciences. New York, McGraw-Hill, 1960.
(H182)

Schimper, Andreas F. W. Plant-geography upon a physiological basis. Oxford, Clarendon Press, 1903. (H183)

6. Soils

Buckman, Harry O., and Brady, Nyle C. The nature and
property of soils. 6th ed. New York, Macmillan, 1960.
(H184)

Bunting, Brian T. The geography of soils. Chicago, Aldine,
1965. (H185)

Donahue, Roy L. Soils: an introduction to soils and plant
growth. 2nd ed. Englewood Cliffs, New Jersey, Prentice-Hall, 1965. (H186)

United States. Department of Agriculture. Soil: the 1957
yearbook of agriculture. Washington, Government Printing
Office, 1957. (H187)

b. Cultural Geography

1. Urban

Chapin, F. Stuart. Urban Land Use Planning. 2nd ed.
Urbana, University of Illinois Press, 1965. (H188)

Gottmann, Jean. Megalopolis: the urbanized north-eastern
seaboard of the United States. Cambridge, M. I. T. Press,
1964. (H189)

Hauser, Philip M., and Schnore, Leo F. (eds.). The study
of urbanization. New York, Wiley, 1965. (H190)

Mayer, Harold M., and Kohn, Clyde F. (eds.). Readings in
urban geography. Chicago, University of Chicago Press,

1959. (H191)

Mumford, Lewis. The city in history: its origins, its
  transformations, and its prospects. New York, Harcourt,
  Brace and World, 1961. (H192)

2. Housing

Abrams, Charles. Man's struggle for shelter in an urbanizing world. Cambridge, M.I.T. Press, 1964. (H193)

American Health Association. Committee on the Hygiene of
  Housing. An appraisal method for measuring the quality
  of housing. New York, American Health Association,
  1945-1950. (H194)

Ettinger, Jan. Towards a habitable world; tasks, problems
  and methods, acceleration. New York, Elsevier, 1960.
  (H195)

Mumford, Lewis. From the ground up. New York, Harcourt, Brace and World, 1956. (H196)

United Nations. Economic Commission for Europe. Techniques for surveying a country's housing situation, including estimating of current and future housing requirements. Geneva, United Nations, 1962. (H197)

Wendt, Paul F. Housing policy: the search for solutions.
  Berkeley, University of California Press, 1962. (H198)

3. Population

Beaujeu-Garnier, J. Geography of population. New York,
  St. Martin's Press, 1966. (H199)

Clarke, John I. Population geography. Oxford, Pergamon
  Press, 1965. (H200)

Hauser, Philip M., and Duncan, Otis. (eds.). The study
  of population; an inventory and appraisal. Chicago, University of Chicago Press, 1959. (H201)

Thompson, Warren S., and Lewis, David T. Population
  problems. 4th ed. New York, McGraw-Hill, 1953.
  (H202)

United Nations. Department of Economic and Social Affairs.
Population Division. Determinants and consequences of
population trends; a summary of the findings of studies
on the relationships between population changes and economic and social conditions. Population Studies Series
No. 17. New York, Columbia University Press, 1954.
(H203)

Zelinsky, Wilbur. A prologue to population geography.
Englewood Cliffs, New Jersey, Prentice-Hall, 1966. (H204)

4. Medical

May, Jacques M. Studies in disease ecology. Studies in
Medical Geography Vol. 2. New York, Hafner, 1961.
(H205)

Stamp, L. Dudley. The geography of life and death. London, Collins, 1964. (H206)

_____. Some aspects of medical geography. University
of London Health Clark Lectures. New York, Oxford
University Press, 1964. (H207)

Tromp, Solco W. Medical biometeorology: weather, climate
and the living organism. New York, Elsevier, 1963.
(H208)

5. Economic

a. General

Alexander, John W. Economic geography. Englewood
Cliffs, New Jersey, Prentice-Hall, 1963. (H209)

Bengtson, Nels A., and Van Royen, William. Fundamentals
of economic geography; an introduction to the study of resources. Englewood Cliffs, New Jersey, Prentice-Hall,
1964. (H210)

Boesch, Hans H. A geography of world economy. Princeton, New Jersey, Van Nostrand, 1964. (H211)

Gregor, Howard F. Environment and economic life. Princeton, New Jersey, Van Nostrand, 1963. (H212)

Yeates, Maurice H. An introduction to quantitative analysis

in economic geography. New York, McGraw-Hill, 1968.
(H213)

Thoman, Richard S. The geography of economic activity;
an introductory world survey. New York, McGraw-Hill,
1962. (H214)

White, Charles L. World economic geography. Belmont,
California, Wadsworth, 1964. (H215)

b. Industrial

Estall, R. C., and Buchanan, Ogilvie R. Industrial activity
and economic geography. New York, Humanities Press,
1961. (H216)

Miller, R. Willard. A geography of manufacturing. Englewood Cliffs, New Jersey, Prentice-Hall, 1962. (H217)

c. Agriculture

Higbee, Edward C. American agriculture; geography, resources, conservation. New York, Wiley, 1958. (H218)

Klages, Karl H. W. Ecological crop geography. New York,
Macmillan, 1962. (H219)

Sauer, Carl O. Agricultural origins and dispersals. New
York, American Geographical Society, 1952. (H220)

United States. Department of Agriculture. Yearbooks of
agriculture. Washington, Government Printing Office,
var. dates. (H221)

Van Royen, William. Atlas of the world's resources. Vol.
1, agricultural resources of the world. New York, Prentice-Hall, 1954. (H222)

d. Transportation

Appleton, J. H. A morphological approach to the geography
of transport. Hull, England, University of Hull, 1965.
(H223)

Daggett, Stuart. Principles of inland transportation. 4th ed.
New York, Harper and Brothers, 1955. (H224)

Garrison, William L. (and others). Studies of highway development and geographic change. Seattle, University of Washington Press, 1959. (H225)

Kansky, Karol J.   Structure of transportation networks; relationships between network geometry and regional characteristics. Department of Geography Research Paper No. 84. Chicago, University of Chicago Press, 1963. (H226)

Ullman, Edward L.   American Commodity flow; a geographic interpretation of rail and water traffic based on principles of spatial interchange.   Seattle, University of Washington Press, 1957. (H227)

6. Land Use and Land Tenure

Chisholm, Michael. Rural settlement and land use.   London, Hutchinson University Library, 1962. (H228)

Lord, Russell.   The care of the earth.   New York, New American Library of World Literature, 1962. (H229)

Marschner, F. J.   Land use and its patterns in the United States. United States Department of Agriculture. Agricultural Handbook No. 153. Washington Government Printing Office, 1959. (H230)

Ottoson, Howard W. (ed.).   Land use policy and problems in the United States.   Lincoln, University of Nebraska Press, 1963. (H231)

United States. Department of Agriculture. A place to live; the 1963 yearbook of the Department of Agriculture. Washington, Government Printing Office, 1963. (H232)

_____. _____. Land; the 1958 yearbook of agriculture. Washington, Government Printing Office, 1958. (H233)

7. Resources

Bateman, Alan M. (ed.).   Economic geology. Urbana, Illinois, Economic Geology Publishing Co., 1955. (H234)

Carlson, Albert S.   Economic geography of industrial materials. New York, Reinhold, 1956. (H235)

Colby, Charles C., and Foster, Alice.   Economic geography;

industries and resources of the commercial world. Boston, Ginn, 1940. (H236)

8. Trade and Commerce

Alexanderson, Gunnar, and Goran, Norstrom. World shipping; an economic geography of ports and seaborne trade. Stockholm, Almquist and Wiksell, 1963. (H237)

Alnwick, Herbert. A geography of commodities. London, Harrap, 1959. (H238)

Chisholm, George G. Handbook of commercial geography. 16th ed. London, Longmans, Green, 1960. (H239)

9. Economic Development

Gill, Richard T. Economic development; past and present. Englewood Cliffs, New Jersey, Prentice-Hall, 1963. (H240)

Higgins, Benjamin H. Economic development; principles, problems, and policies. New York, Norton, 1959. (H241)

Jackson, Barbara (Ward). The rich nations and the poor nations. New York, Norton, 1962. (H242)

Kindleberger, Charles P. Economic development. 2nd ed. New York, McGraw-Hill, 1965. (H243)

Shannon, Lyle W. Underdeveloped areas, a book of readings and research. New York, Harper, 1957. (H244)

10. Political

Cohen, Saul B. Geography and politics in a world divided. New York, Random House, 1963. (H245)

De Blij, Harm J. Systematic political geography. New York, Wiley, 1967. (H246)

Jackson, W. A. Douglas (ed.). Politics and geographic relationships: readings on the nature of political geography. Englewood Cliffs, New Jersey, Prentice-Hall, Inc., 1964. (H247)

Pounds, Norman J. G. Political geography. New York,

McGraw-Hill, 1963. (H248)

Weigert, Hans W., and Stefansson, Viljhalmur (eds.). New compass of the world; a symposium on political geography. New York, Macmillan, 1949. (H249)

_____ (and others). Principles of political geography. New York, Appleton-Century-Crofts, 1957. (H250)

11. Military

Mackinder, Halford J. Britain and the British seas. New York, Appleton, 1902. (H251)

Mahan, Alfred T. The influence of sea power upon history 1660-1783. Boston, Little, Brown, 1890. (H252)

Peltier, Louis, and Pearcy, G. Etzel. Military geography. Princeton, New Jersey, Van Nostrand, 1966. (H253)

United States. Air Force. Reserve Officers Training Corps. Military aspects of world political geography. Alabama, Maxwell Air Force Base, 1959. (H254)

12. Historical

Brown, Ralph H. Historical geography of the United States. New York, Harcourt, Brace and World, 1948. (H255)

Butzer, Karl W. Environment and archeology; an introduction to Pleistocene geography. Chicago, Aldine, 1964. (H256)

Childe, V. Gordon. Man makes himself. New York, New American Library, 1951. (H257)

East, William G. An historical geography of Europe. London, Methuen, 1935. (H258)

Paullin, Charles O. (ed.). Atlas of the historical geography of the United States. Washington, Carnegie Institution, 1932. (H259)

Sykes, Percy. A history of exploration from the earliest times to the present day. 3rd ed. New York, Macmillan, 1950. (H260)

Thomas, William L., Jr. (ed.). Man's role in changing the
   face of the earth. Chicago, University of Chicago Press,
   1956. (H261)

2. Regional Geography

a. Anglo-America

American Geographical Society. Readings in geography of
   North America; a selection of articles from the Geograph-
   ical Review. Reprint Series No. 5. New York, American
   Geographical Society, 1952. (H262)

Paterson, John H. North America; a regional geography.
   London, Oxford University Press, 1965. (H263)

Putman, D. F., and Kerr, D. P. A regional geography of
   Canada. Toronto, J. M. Dent and Sons, 1956. (H264)

Smith, J. Russell, and Phillips, M. Ogden. North America.
   New York, Harcourt, Brace and World, 1942. (H265)

White, C. Langdon, Foscue, Edwin, and McKnight, Tom.
   Regional geography of Anglo-America. Englewood Cliffs,
   New Jersey, Prentice-Hall, 1964. (H266)

Mead, William R., and Brown, E. H. The United States
   and Canada; a regional geography. London, Hutchinson
   Educational Co., 1962. (H267)

b. Latin America

James, Preston E. Latin America. 4th ed. New York,
   Odyssey, 1969. (H268)

Platt, Robert S. Latin America; countrysides and united
   regions. New York, McGraw-Hill, 1943. (H269)

Steward, Julian H. (ed.). Handbook of South American Indians. Washington, Government Printing Office, 1946-
   1957. (H270)

Wauchope, Robert (ed.). Handbook of Middle American Indians. 8 vols. Austin, University of Texas Press, 1964-
   1969. (H271)

Wagley, Charles (ed.). Social science research on Latin
  America. New York, Columbia University Press, 1964.
  (H272)

c. Europe

Dewhurst, J. Fredric, Coppock, John O., Yates, P. Lamar-
  tine, and Associates. Europe's needs and resources;
  trends and prospects in eighteen countries. New York,
  Twentieth Century Fund, 1961. (H273)

Dickinson, Robert E. The West European city. London,
  Routledge and Kegan Paul, 1951. (H274)

Gottmann, Jean. A geography of Europe. 2nd ed. New
  York, Holt, 1954. (H275)

Hoffman, George W. (ed.). A geography of Europe, including Asiatic U.S.S.R. 2nd ed. New York, Ronald, 1961.
  (H276)

Pounds, Norman J. G. Europe and the Soviet Union, 2nd
  ed. New York, McGraw-Hill, 1966. (H277)

d. Soviet Union

Mellor, Roy E. H. Geography of the U.S.S.R. New York,
  St. Martins, 1965. (H278)

Jorre, Georges. The Soviet Union, the land and its people.
  Translated by E. D. Laborde. 2nd ed. New York, Wiley,
  1961. (H279)

Cole, John P., and German, Frank. A geography of the
  U.S.S.R.; the background to a planned economy. London,
  Butterworth, 1961. (H280)

Cressey, George B. Soviet potentials: a geographic appraisal. Syracuse, Syracuse University Press, 1962.
  (H281)

e. Africa

Hailey, William M. H. (Baron). An African survey; a study
  of the problems arising in Africa south of the Sahara.
  London, Oxford University Press, 1957. (H282)

Hance, William A. Geography of modern Africa. New York, Columbia University Press, 1964. (H283)

Kimble, George H. T. Tropical Africa. New York, Twentieth Century Fund, 1960. (H284)

Lystad, Robert A. The African world; a survey of social research. New York, Praeger, 1965. (H285)

Stamp, L. Dudley. Africa, a study in tropical development. 2nd ed. New York, Wiley, 1964. (H286)

f. Asia

Cressey, George B. Asia's land and peoples; a geography of one-third of the earth and two-thirds of its people. 3rd ed. New York, McGraw-Hill, 1963. (H287)

East, W. Gordon, and Spate, Oscar H. K. (eds.). The changing map of Asia. 4th ed. New York, Dutton, 1962. (H288)

Ginsburg, Norton S. (and others). The pattern of Asia. Englewood Cliffs, New Jersey, Prentice-Hall, 1958. (H289)

Spencer, Joseph E. Asia, east by south, a cultural geography. New York, Wiley, 1954. (H290)

Stamp, L. Dudley. Asia, a regional and economic geography. 11th ed. New York, Dutton, 1962. (H291)

g. Australia - New Zealand - Pacific

Cumberland, Kenneth B. Southwest Pacific; a geography of Australia, New Zealand, and their Pacific island neighbors. New York, McGraw-Hill, 1956. (H292)

_____, and Fox, James W. New Zealand, a regional view. 2nd ed. Christchurch, Witcombe and Tombs, 1963. (H293)

Robinson, Kathleen W. Australia, New Zealand and the Southwest Pacific. New York, London House and Maxwell, 1962. (H294)

Taylor, T. Griffith. Australia; a study of warm environments and their effect on British settlement. 7th ed. London, Methuen, 1959. (H295)

Younger, R. M. The changing world of Australia. New York, Franklin Watts, 1963. (H296)

h. Polar Areas

Baird, Patrick D. The Polar World. London, Longmans, Green, 1964. (H297)

Gould, Laurence M. The Polar regions in the relation to human affairs. Bowman Memorial Lectures, No. 4. New York, American Geographical Society, 1958. (H298)

Hatherton, Trevor (ed.). Antarctica. New York, Praeger, 1965. (H299)

Kimble, George H. T., and Good, Dorothy (eds.). Geography of the northlands. American Geographical Society Special Publication No. 2. New York, American Geographical Society and John Wiley, 1955. (H300)

Nanson, Fridtjof. In northern mists; Arctic exploration in early times. 2 vols. New York, Frederick A. Stokes, 1911. (H301)

3. Geographical Techniques

a. Cartography

Boggs, Samuel W., and Lewis, Dorothy C. The classification and cataloging of maps and atlases. New York, Special Libraries Association, 1945. (H302)

Brown, Lloyd A. The story of maps. Boston, Little, Brown, 1949. (H303)

Greenhood, David. Mapping. Phoenix Science Series No. 521. Chicago, University of Chicago Press, 1964. (H304)

Monkhouse, F. J., and Wilkinson, H. R. Maps and diagrams; their compilation and construction. 2nd ed. New York, Dutton, 1963. (H305)

Raisz, Erwin J. General cartography. 2nd ed. New York,

McGraw-Hill, 1948. (H306)

b. Photo Interpretation and Remote Sensing

American Society of Photogrammetry. Manual of photographic
    interpretation. Washington, American Society of Photo-
    grammetry, 1960. (H307)

Lueder, Donald R. Aerial photographic interpretation; prin-
    ciples and application. New York, McGraw-Hill, 1961.
    (H308)

Miller, Victor C. Photogeology. New York, McGraw-Hill,
    1961. (H309)

Smith, Harold T. U. Aerial photographs and their applica-
    tions. New York, Appleton-Century-Crofts, 1943. (H310)

Spurr, Stephen H. Photogrammetry and photo-interpretation,
    with a section on applications to forestry. 2nd ed. New
    York, Ronald, 1960. (H311)

c. Geography Teaching Aids

American Association of Petroleum Geologists. Committee
    on Preparation of Projection Slides. AAPG slide manual;
    a guide to the preparation and use of projection slides.
    Tulsa, Oklahoma, American Association of Petroleum Ge-
    ologists, 1960. (H312)

American Geological Institute. Committee on Education. Out-
    standing aerial photographs in North America. Washing-
    ton, American Geological Institute, 1951. (H313)

Gabler, Robert E. A handbook for geography teachers. Nor-
    mal, Illinois, National Council for Geographic Education,
    1966. (H314)

Heller, Robert L. (ed.). Geology and Earth Sciences source-
    book for elementary and secondary schools. New York,
    Holt, Rinehart and Winston, 1962. (H315)

4. History and Philosophy of Geography

a. General Readings

Ackerman, Edward A. Geography as a fundamental research

discipline. Geography Research Paper No. 53. Chicago, University of Chicago Press, 1958. (H316)

Bowman, Isiah. Geography in relation to the social sciences. New York, Scribner's, 1934. (H317)

Broek, Jan O. M. Geography; its scope and spirit. Columbus, Ohio, Charles E. Merrill Books, Inc., 1965. (H318)

Brunhes, Jean. Human geography; an attempt at a positive classification, principles and examples. Chicago, Rand McNally, 1920. (H319)

Bunge, William W. Theoretical geography. Lund, Sweden, C. W. K. Gleerup, 1962. (H320)

Dickinson, Robert E., and Howarth, Osbert J. R. The making of geography. Oxford, Clarendon Press, 1933. (H321)

Freeman, T. W. A hundred years of geography. Chicago, Aldine, 1961. (H322)

Hartshorne, Richard. Perspective on the nature of geography. Chicago, Rand McNally, 1959. (H323)

_____. The nature of geography: a critical survey of current thought in the light of the past. Lancaster, Pennsylvania, Association of American Geographers, 1939. (H324)

Harvey, David. Explanation in geography. London, Edward Arnold, 1969. (H325)

James, Preston E., and Jones Clarence F. (ed.). American geography: inventory and prospect. Syracuse, New York, Syracuse University Press, 1954. (H326)

National Academy of Sciences. National Research Council. Earth Sciences Division. The Science of geography: report of the ad hoc committee on geography. Washington, National Academy of Sciences--National Research Council, 1965. (Publication no. 1277.) (H327)

Stamp, Dudley. Applied geography. Harmondsworth, Middlesex, Penguin Books, 1960. (H328)

Taylor, Griffith (ed.). Geography in the twentieth century;

112 Library Research in Geography

a study of growth, fields, techniques, aims, and trends.
3rd ed. enl. New York, Philosophical Library, 1957.
(H329)

Wooldridge, Sidney W., and East, William G. The spirit
and purpose of geography. rev. ed. New York, Putnam,
1967. (H330)

b. Some Primary Readings on the History,
   Philosophy, and Methodology of Geography

Ackerman, E. A. "Geographic training, wartime research,
and immediate professional objectives." Annals. 35:121-
143, 1945. (H331)

_____. "Regional research--emerging concepts and tech-
niques in the field of geography." Economic Geography.
29:189-197, 1953. (H332)

_____. "The Köppen classification of climates in North
America." Geographical Review. 31:105-111, 1941.
(H333)

_____. "Where is a research frontier?" Annals. 53:429-
440, 1963. (H334)

Alexander, J. W. "Geography: as some others see it."
The Professional Geographer. 11:2-5, 1959. (H335)

Bachi, R. "Standard distance measures and related methods
for spatial analysis." Regional Science Association, Papers and Proceedings. 10:83-132, 1963. (H336)

Baker, O. E. "Population, food supply, and American agriculture." Geographical Review. 18:353-373, 1928.
(H337)

Ballabon, M. B. "Putting the 'economic' into economic geography." Economic Geography. 33:217-223, 1957. (H338)

Barrows, H. H. "Geography as human ecology." Annals.
13:1-14, 1923. (H339)

Baum, W. A. and Court, A. "Research status and needs in
microclimatology." Transactions of the American Geophysical Union. 30:488-493, 1949. (H340)

Subject Bibliographies   113

Beckmann, M. J. "Some reflections on Lösch's theory of
   location." Regional Science Association, Papers and Proceedings. 1:N1-N9, 1955. (H341)

Bennett, A. S. "Some aspects of preparing questionnaires."
   Journal of Marketing. 10:175-179, 1945. (H342)

Berry, B. J. L. "A method for deriving multifactor uniform
   regions." Przeglad Geograficzny. 33:263-282, 1961.
   (H343)

_____. "Approaches to regional analysis: a synthesis." An-
   nals. 54:2-11, 1964. (H344)

_____. "Further comments concerning 'geographic' and 'eco-
   nomic' geography." The Professional Geographer, 12:11-
   12, 1959. (H345)

_____, and Garrison, W. L. "A note on central place theory
   and the range of a good." Economic Geography. 34:304-
   311, 1958. (H346)

Birch, J. W. "A note on the sample-farm survey and its
   use as a basis for generalized mapping." Economic Geography. 36:254-259, 1960. (H347)

Blaut, J. M. "Microgeographic sampling: a quantitative
   approach to regional agricultural geography." Economic
   Geography. 35:79-88, 1959. (H348)

_____. "Space and process." Professional Geographer. 13:
   1-7, 1961. (H349)

Blumenfeld, H. "On the concentric-circle theory of urban
   growth." Land Economics. 25:209-212, 1949. (H350)

Borchert, J. R. "A statement favoring support of the term
   'geography'." The Professional Geographer. 12:14-16,
   1960. (H351)

Bowman, Isiah. "Geography: interpretations." Geographical
   Review. 39:355-370, 1949. (H352)

_____. "Geography vs. geopolitics." Geographical Review.
   32:646-658, 1942. (H353)

_____. "The scientific study of settlement." Geographical

Review. 16:647-653, 1926. (H354)

Brigham, A. P. "Problems of geographical influence." Annals. 5:3-25, 1915. (H355)

Broek, Jan O. M. "Discourse on economic geography."
Geographical Review. 31:663-674, 1941. (H356)

_____. "The relations between history and geography."
Pacific Historical Review. 10:321-325, 1941. (H357)

_____, and Innis, Harold A. "A discussion of geography and
nationalism." Geographical Review. 35:301-311, 1945.
(H358)

Brookfield, H. C. "Questions on the human frontiers of geography." Economic Geography. 40:283-303, 1964.
(H359)

Brown, R. H. "The American geographies of Jedidiah
Morse." Annals. 31:145-217, 1941. (H360)

Brown, W. J. "A lawyer looks at the name geography."
Professional Geographer. 13:16-19, 1961. (H361)

Bryan, K. "The place of geomorphology in the geographic
sciences." Annals. 40:196-208, 1950. (H362)

Buchanan, R. O. "Some reflections on agricultural geography." Geography. 44:1-13, 1959. (H363)

Bunge, W. "Geographical dialectics." Professional Geographer. 16:28-29, 1964. (H364)

_____. "Locations are not unique." Annals. 56:375-376,
1966. (H365)

Burker, L. V. "Geography and space." Geographical Review. 49:305-314, 1959. (H366)

Burton, I. "The quantitative revolution and theoretical geography." The Canadian Geographer. 7:151-168, 1963.
(H367)

Calef, W. C. "Methodology through the looking glass."
Geographical Review. 47:428-431, 1957. (H368)

Carol, H. "Geography of the future." The Professional
Geographer. 13:14-18, 1957. (H369)

Carrothers, G. P. "An historic review of the gravity and
potential concepts of human interaction." Journal of the
American Institute of Planners. 22:94-102, 1956. (H370)

Carter, H. C. "Representative farms--guides for decision
making." Journal of Farm Economics. 45:1449-1455,
1963. (H371)

Chisholm, M. "Problems in the classification and use of
the farming type region." Transactions of the Institute
of British Geographers. 35:91-103, 1964. (H372)

Chorley, R. J. "Geography and analogue theory." Annals.
54:127-137, 1964. (H373)

Clark, A. H. "What geographers did: a review." Economic Geography. 44:83-86, 1968. (H374)

Cochran, W. G., Mosteller, F., and Tukey, J. W. "Principles of sampling." Journal of the American Statistical
Association. 49:13-35, 1954. (H375)

Colby, C. C. "Centrifugal and centripetal forces in urban
geography." Annals. 23:1-20, 1933. (H376)

_____. "Changing currents of geographic thought in Ameri-
ca." Annals. 26:1-37, 1936. (H377)

_____. "The railway traverse as an aid in reconnaissance."
Annals. 23:157-164, 1933. (H378)

Crowe, P. R. "On progress in geography." Scottish Geographical Magazine. 54:1-19, 1938. (H379)

Curry, L. "The random spatial economy: an exploration in
settlement theory." Annals. 54:138-146, 1964. (H380)

Davies, J. L. "Aim and method in zoogeography." Geographical Review. 51:412-417, 1961. (H381)

Davis, W. M. "An inductive study of the content of geography." Bulletin of the American Geographical Society.
38:67-84, 1906. (H382)

_____. "The geographical cycle." Geographical Journal. 14:481-504, 1899. (H383)

_____. "The principles of geographic description." Annals. 1:61-105, 1915. (H384)

_____. "The progress of geography in the United States." Annals. 14:159-215, 1924. (H385)

_____. "A retrospect of geography." Annals. 22:211-230, 1932. (H386)

DeGeer, S. "On the definition, method and classification of geography." Geografiska Annaler. 5:1-37, 1923. (H387)

Dodge, S. D. "Geography: What to call it." The Professional Geographer. 12:13-14, 1960. (H388)

Dryer, C. R. "Genetic geography: the development of the geographic sense and concept." Annals. 10:3-16, 1920. (H389)

Dunn, E. S. "The market potential concept and the analysis of location." Regional Science Association, Papers and Proceedings. 2:183-194, 1956. (H390)

Egler, F. E. "Vegetation as an object of study." Philosophy of Science. 9:245-260, 1942. (H391)

Fenneman, N. M. "Physiographic divisions of the United States." Annals. 18:261-353, 1928. (H392)

_____. "The circumference of geography." Annals. 13:1-14, 1923. (H393)

Finch, V. C. "Geographical science and social philosophy." Annals. 29:1-28, 1939. (H394)

_____. "Training for research in economic geography." Annals. 34:207-215, 1944. (H395)

Fisher, C. A. "The status of American geography: a review." Economic Geography. 31:87-90, 1955. (H396)

Fitzgerald, W. "Regional concept in geography: recognition of humanistic interests." Nature. 152:740-741, 1943. (H397)

Fleure, H. J. "Geographical thought in the changing world." Annals. 34:515-528, 1944. (H398)

Floyd, B. N. "The pleasures ahead: a geographic meditation." The Professional Geographer. 51:1-4, 1962. (H399)

_____. "Toward a more literary geography." The Professional Geographer. 13:7-11, 1961. (H400)

_____. "Quantification--a geographic deviation?" The Professional Geographer. 15:15-17, 1963. (H401)

Forgotson, J. M. "Review and classification of quantitative mapping techniques." Bulletin of the American Association of Petroleum Geologists. 44:83-100, 1960. (H402)

Garrison, W. L. "Applicability of statistical inference to geographical research." Geographical Review. 46:427-429, 1956. (H403)

_____, and Marble, D. F. "The spatial structure of agricultural activities." Annals. 47:137-144, 1957. (H404)

Gauld, W. A. "Towards a new geography." Nature. 147:546-548, 1941. (H405)

Gilbert, E. W. "The idea of the region." Geography. 45:157-175, 1960. (H406)

Gottman, J. "Vauban and modern geography." Annals. 34:120-128, 1944. (H407)

Gould, P. R. "Man against his environment: a game-theoretic framework." Annals. 53:290-297, 1963. (H408)

Grotewald, A. "Von Thunen in retrospect." Economic Geography. 35:346-355, 1959. (H409)

Hagood, M. J. "Statistical methods for delineation of regions applied to data on agriculture and population." Social Forces. 21:288-297, 1943. (H410)

Hall, R. B. "The geographic region: a resumé." Annals. 25:122-136, 1935. (H411)

Hanson, R. M. "What geography do you want?" Journal of Geography. 58:33-43, 1959. (H412)

Harris, B. F. D. "The claims of geography to be considered as a science and consequent implications as to
methods of teaching the subject." Geography. 20:38-46,
1935. (H413)

Hart, J. F. "Central tendency in areal distributions." Economic Geography. 30:48-59, 1954. (H414)

_____. "'Exceptionalism in geography' re-examined." An-
nals. 45:205-244, 1955. (H415)

Hartshorne, R. "Location as a factor in geography." Annals. 17:92-99, 1927. (H416)

_____. "On the mores of methodological discussion in Amer-
ican geography." Annals. 18:113-125, 1948. (H417)

_____. "Recent developments in political geography." Amer-
ican Political Science Review. 29:784-804; 943-966, 1935.
(H418)

_____. "The content of geography as a science of space
from Kant and Humboldt to Hettner." Annals. 48:97-108,
1958. (H419)

_____. "The economic geography of plant location." Annals
of Real Estate Practice. 6:40-76, 1926. (H420)

_____. "The politico-geographic pattern of the world." An-
nals of the American Academy of Political and Social Science. 218:45-57, 1941. (H421)

_____. "What do we mean by region?" Annals. 48:268,
1958. (H422)

_____, and Dicken, S. N. "A classification of the agri-
cultural regions of Europe and North America on a
uniform statistical basis." Annals. 25:99-120, 1935.
(H423)

Hartsman, G. W. "The central business district--A study
in urban geography." Economic Geography. 26:237-244,
1950. (H424)

Harris, C. D. "A functional classification of cities in the
United States." Geographical Review. 33:86-99, 1943.
(H425)

Harvey, D. W. "Theoretical concepts and the analysis of agricultural land-use patterns in geography." Annals. 56:361-374, 1966. (H426)

Helburn, N. "The bases for a classification of world agriculture." Professional Geographer. 9:2-7, 1957. (H427)

Heller, C. F. "The use of model farms in agricultural geography." Professional Geographer. 16:20-23, 1964. (H428)

Herbertson, A. J. "The major natural regions: An essay in systematic geography." Geographical Journal. 25: 300-321, 1905. (H429)

Hewes, Leslie. "On the common responsibility of economic and cultural geography." Economic Geography. 42:94, 1966. (H430)

Hudson, G. D. "Methods employed by geographers in regional surveys." Economic Geography. 12:98-104, 1936. (H431)

_____. "The unit area method of land classification." Annals. 26:99-112, 1936. (H432)

Huntington, Ellsworth. "What next in geography?" Journal of Geography. 41:1-9, 1942. (H433)

Isard, W. "Regional science: the concept of the region and regional structure." Papers and Proceedings of the Regional Science Association. 2:13-26, 1956. (H434)

James, Preston E. "On the origin and persistence of error in geography." Annals. 57:1-24, 1967. (H435)

_____. "The geography of the oceans: A review of the work of Gerhard Schott." Geographical Review. 26:664-669, 1936. (H436)

_____. "Toward a further understanding of the regional concept." Annals. 42:195-222, 1952. (H437)

Jefferson, M. "The civilizing rails." Economic Geography. 4:217-231, 1928. (H438)

_____. "The law of the primate city." Geographical Review.

29:226-232, 1939. (H439)

Johnson, D. "The geographic prospect." Annals. 19:167-231, 1929. (H440)

Johnson, L. J. "Some thoughts on geography of the future." The Professional Geographer. 13:30-32, 1961. (H441)

Jones, Emrys. "Cause and effect in human geography." Annals. 46:369, 1956. (H442)

Jones, S. B. "The economic geography of atomic energy, A review article." Economic Geography. 27:268-274, 1951. (H443)

_____. "The enjoyment of geography." Geographical Review. 42:543-550, 1952. (H444)

Jones, W. D. and Finch, V. C. "Detailed field mapping in the study of the economic geography of an agricultural area." Annals. 15:148-157, 1925. (H445)

_____, and Sauer, C. O. "Outline for field work in geography." Bulletin of the American Geographical Society. 47:520-525, 1915. (H446)

Kesseli, J. E. "Use of air photographs by geographers." Photogrammetric Engineering. 18:737-741, 1952. (H447)

King, L. J. "A note on theory and reality." The Professional Geographer. 12:406, 1960. (H448)

_____. "The analysis of spatial form and its relation to geographic theory." Annals. 59:573-596, 1969. (H449)

Kirk, W. "Problems in geography." Geography. 48:357-371, 1963. (H450)

Klimm, Lester E. "Mere description." Economic Geography. 35:1, 1959. (H451)

Kohn, C. F. "The use of aerial photographs in the geographical analysis of rural settlements." Photogrammetric Engineering. 17:759-771, 1951. (H452)

_____. "The 1960's: a decade of progress in geographical research and instruction." Annals. 60:211-219, 1970. (H453)

Kollmorgen, W. M. "Crucial deficiencies of regionalism." American Economic Review: Papers and Proceedings. 35:377-389, 1945. (H454)

Küchler, A. W. "A physiognomic classification of vegetation." Annals. 39:201-210, 1949. (H455)

_____. "The relation between classifying and mapping vegetation." Ecology. 32:275-283, 1951. (H456)

LaValle, P., McConnell, H. and Brown, R. G. "Certain aspects of the expansion of quantitative methodology in American geography." Annals. 57:423-436, 1967. (H457)

Leighly, John. "Problems and trends in American geography." Geographical Review. 59:161, 1969. (H458)

_____. "Some comments on contemporary geographic methods." Annals. 27:125-141, 1937. (H459)

Lewis, P. W. "Three related problems in the formulation of laws in geography." The Professional Geographer. 17:24-27, 1965. (H460)

Lewthwaite, G. R. "Environmentalism and determinism: a study in confusion." Annals. 53:604-605, 1963. (H461)

Lively, C. E. and Gregory, C. C. "The rural socio-cultural area of a field for research." Rural Sociology. 19:21-31, 1954. (H462)

Lösch, A. "The nature of economic regions." The Southern Economic Journal. 5:71-78, 1938. (H463)

Lowenthal, D. "Geography, experience and imagination: toward a geographic epistemology." Annals. 51:241-260, 1961. (H464)

_____. "George Perkins Marsh and the American geographical tradition." Geographical Review. 43:207-213, 1953. (H465)

Lucien, Febvre. "A geographical introduction to history." Geographical Review. 13:147-148, 1923. (H466)

Lukermann, F. "On explanation, model and description." The Professional Geographer. 12:1-2, 1958. (H467)

_____. "The role of theory in geographic inquiry." The Professional Geographer. 13:1-5, 1961. (H468)

_____. "Towards more geographic economic geography." The Professional Geographer. 10:2-13, 1958. (H469)

MacFadden, C. H. "The uses of aerial photographs in geographic research." Photogrammetric Engineering. 18: 732-737, 1952. (H470)

Mackay, J. R. "Chi-square as a tool for regional studies." Annals. 48:164, 1958. (H471)

Mackinder, H. J. "The geographical pivot of history." Geographical Journal. 23:421-437, 1904. (H472)

Malin, J. C. "Ecology and history." Scientific Monthly. 70:295-298, 1950. (H473)

Mattice, W. A. "The future of agricultural meteorology." Monthly Weather Review. 59:274-275, 1931. (H474)

May, J. M. "Medical geography: Its methods and objectives." Geographical Review. 40:9-41, 1950. (H475)

Mayer, H. M. "City retail structure." Economic Geography. 13:425-428, 1937. (H476)

_____. "Geography and urban planning: theory and application." Economic Geography. 42:282, 1966. (H477)

McCarty, H. H. "A functional analysis of population distribution." Geographical Review. 32:282-293, 1942. (H478)

_____. "Use of certain statistical procedures in geographical-analysis." Annals. 46:263, 1956. (H479)

McMurry, K. C. "The use of land for recreation." Annals. 20:7-20, 1930. (H480)

McNee, Robert B. "The changing relationships of economics and economic geography." Economic Geography. 35:189-198, 1959. (H481)

Meigs, P. "Water problems in the United States." Geographical Review. 42:346-366, 1952. (H482)

Morgan, W. B. and Moss, R. P. "Geography and ecology: the concept of the community and its relationship to environment." Annals. 55:339-350, 1965. (H483)

Murphey, R. E. "Geography: how adequate is the term?" The Professional Geographer. 12:8-10, 1960. (H484)

Murphy, R. C. "Animal geography: A review." Geographical Review. 28:140-144, 1938. (H485)

Parkins, A. E. "The geography of American geographers." Journal of Geography. 33:221-230, 1934. (H486)

Penrose, E. F. "The place of transport in economic and political geography." Transport and Communications Review. 5:1-8, 1952. (H487)

Philbrick, A. K. "Principles of areal functional organization in regional human geography." Economic Geography. 33: 299-336, 1957. (H488)

Platt, R. S. "Determinism in geography." Annals. 38: 126-132, 1948. (H489)

_____. "Environmentalism versus geography." American Journal of Sociology. 53:351-358, 1948. (H490)

_____. "Field approach to regions." Annals. 25:153-174, 1935. (H491)

_____. "Review of regional geography." Annals. 47:187-190, 1957. (H492)

Poole, Sidman P. "Post-war geography; an editorial." Economic Geography. 19:320-322, 1943. (H493)

Porter, P. W. "Does geography need a social physic?" Annals. 50:341, 1960. (H494)

Quant, R. E. "Models of transportation and optimal network construction." Journal of Regional Science. 2:27-45, 1960. (H495)

Raisz, E. "Landform, landscape, land-use, and land-type maps." Journal of Geography. 45:85-90, 1946. (H496)

Raup, H. M. "Trends in the development of geographic botany." Annals. 32:319-354, 1942. (H497)

Reeds, L. G. "Agricultural geography: progress and prospects." Canadian Geographer. 8:51-63, 1964. (H498)

Renner, G. T. "Geography of industrial localization." Economic Geography. 23:167-189, 1947. (H499)

Reynolds, R. B. "Statistical methods in geographical research." Geographical Review. 46:129-132, 1956. (H500)

Robinson, A. H. "An analytical approach to map projections." Annals. 39:283-290, 1949. (H501)

_____, and Bryson, R. A. "A method for describing quantitatively the correspondence of geographical distributions." Annals. 47:379-391, 1957. (H502)

Roorback, G. B. "The trend of modern geography." Bulletin of the American Geographical Society. 46:801-816, 1914. (H503)

Russell, R. J. "Geographical geomorphology." Annals. 39:1-11, 1949. (H504)

Sauer, C. O. "Foreword to historical geography." Annals. 31:1-24, 1941. (H505)

Schaefer, F. K. "Exceptionalism in geography: a methodological examination." Annals. 43:226-249, 1953. (H506)

Schmid, C. F. and MacCannell, E. H. "Basic problems, techniques, and theory of isopleth mapping." Journal of the American Statistical Association. 50:220-239, 1955. (H507)

Schnore, L. F. "Geography and human ecology." Economic Geography. 37:207-217, 1961. (H508)

Schoy, C. "The geography of the Moslems of the Middle Ages." Geographical Review. 14:257-269, 1924. (H509)

Schulte, O. W. "The use of panchromatic, infrared, and color aerial photography in the study of plant distribution." Photogrammetric Engineering. 17:688-714, 1951. (H510)

Shreve, F. "A map of the vegetation of the United States."

Geographical Review. 3:119-125, 1917. (H511)

_____. "The desert vegetation of North America." Botanical Review. 8:195-246, 1942. (H512)

Sidall, William R. "Two kinds of geography." Economic Geography. 37:188, 1961. (H513)

Smith, H. T. U. "Aerial photographs in geomorphic studies." Journal of Geomorphology. 4:171-205, 1941. (H514)

Smith, W. "The location of industry." Institute of British Geographers Publication. 21:1-18, 1955. (H515)

Spate, O. H. K. "Toynbee and Huntington: a study in determinism." Geographical Journal. 118:4, 1952. (H516)

_____. "The end of an old song?--the determinism-possibilism problem." Geographical Review. 48:280-282, 1958. (H517)

_____. "Quantity and quality in geography." Annals. 50: 377-394, 1960. (H518)

Spencer, J. E. and Horvath, R. J. "How does an agricultural region originate?" Annals. 53:74-92, 1963. (H519)

Spykman, N. J. "Geography and foreign policy." American Political Science Review. 32:28-50; 213-236, 1938. (H520)

Stamp, L. D. "Major natural regions: Herbertson after fifty years." Geography. 42:201-216, 1957. (H521)

Stewart, C. T., Jr. "The size and spacing of cities." Geographical Review. 48:222-245, 1958. (H522)

Stewart, J. Q. "Empirical mathematical rules concerning the distribution and equilibrium of population." Geographical Review. 37:461-485, 1947. (H523)

_____, and Warntz, W. "Macrogeography and social science." Geographical Review. 48:167-184, 1958. (H524)

Stokes, G. A. "The aerial photograph: A key to the cultural landscape." Journal of Geography. 49:32-40, 1950. (H525)

Stone, K. H. "Geographical air-photo interpretation." Photo-grammetric Engineering. 17:754-759, 1951. (H526)

———. "World air photo coverage for geographic research." Annals. 43:193, 1953. (H527)

Strahler, A. N. "Statistical analysis in geomorphic research." Journal of Geology. 62:1-25, 1954. (H528)

Taylor, E. G. R. "Geography in world peace." Geographical Review. 38:132-141, 1948. (H529)

Taylor, Griffith. "Geography, the correlative science." Canadian Journal of Economics and Political Science. 1: 535-550, 1935. (H530)

Teggert, F. J. "Geography as an aid to statecraft; an appreciation of Mackinder's 'Democratic ideals and reality'." Geographical Review. 8:365-374, 1919. (H531)

Thoman, R. S. "Some comments on the 'Science of geography'." The Professional Geographer. 17:8-10, 1965. (H532)

Thomas, E. N. "Towards an expanded central place model." Geographical Review. 51:400-411, 1961. (H533)

Thompson, K. "Geography--a problem in nomenclature." The Professional Geographer. 12:4-6, 1960. (H534)

Thornthwaite, C. W. "An approach toward a rational classification of climate." Geographical Review. 38:55-94, 1948. (H535)

———. "Problems in the classification of climates." Geographical Review. 33:233-255, 1943. (H536)

———. "The climates of North America according to a new classification." Geographical Review. 21:633-655, 1931. (H537)

———. "The task ahead." Annals. 51:345-356, 1961. (H538)

Tower, W. S. "Scientific geography: The relation of its content to its subdivisions." Bulletin of the American Geographical Society. 42:801-825, 1910. (H539)

Trewartha, G. T. "The case for population geography." Annals. 43:71-97, 1953. (H540)

Troll, C. "Geographic science in Germany during the period 1933-1945: a critique and justification." Annals. 39:128-135, 1949. (H541)

Ullman, E. L. "A theory of location for cities." American Journal of Sociology. 46:853-864, 1940-1941. (H542)

_____. "Geography as spatial interaction." Annals. 44: 283f, 1954. (H543)

_____. "Human geography and area research." Annals. 43: 54-66, 1953. (H544)

Vance, R. B. "The concept of the region." Social Forces. 8:208-218, 1929. (H545)

Van Cleef, E. "Geography as an earth science." The Professional Geographer. 12:8-11, 1960. (H546)

_____. "Confusion or revolution?" The Professional Geographer. 16:1-4, 1964. (H547)

_____. "Area differentiation and the 'Science of geography'." Science. 115:654-655, 1953. (H548)

_____. "Must geographers apologize?" Annals. 45:105-108, 1955. (H549)

_____. "Philosophy of geography and geographical regions." Geographical Review. 22:188-194, 1931. (H550)

Valvanis, S. "Lösch on location." American Economic Review. 45:637-644, 1955. (H551)

Visher, S. S. "An American view of geography." The Geographical Teacher. 12:92-95, 1923. (H552)

_____. "Modern geography: its aspects, aims and methods." The Educational Review. 65:295-298, 1923. (H553)

Wagner, Philip L. "Culture and geography, U.S.A." Economic Geography. 42:188, 1966. (H554)

_____. "Remarks on the geography of language." Geographical Review. 48:86-97, 1958. (H555)

Warntz, W. "Contributions toward a macroeconomic geography: a review." Geographical Review. 47:420-424, 1957. (H556)

Watson, J. W. "Geography: a discipline in distance." Scottish Geographical Magazine. 71:1-13, 1955. (H557)

Webb, Martyn J. "Economic geography: a framework for a disciplinary definition." Economic Geography. 37:254-257, 1961. (H558)

Whitaker, J. R. "Way lies open." Annals. 19:162-165, 1929. (H559)

Whitbeck, H. "Fact and fiction in geography by natural regions." Journal of Geography. 22:86-94, 1923. (H560)

Whittlesey, D. S. "Major agricultural regions of the earth." Annals. 26:199-240, 1936. (H561)

_____. "Political geography: A complex aspect of geography." Education. 50:293-298, 1935. (H562)

_____. "Sequent occupance." Annals. 19:162-165, 1929. (H563)

_____. "The horizon of geography." Annals. 35:1-36, 1945. (H564)

Wolpert, J. "The decision process in spatial context." Annals. 54:537-558, 1964. (H565)

Wood, W. F. "Use of stratified random samples in land use study." Annals. 45:350-367, 1955. (H566)

Wright, John K. "Crossbreeding geographical quantities." Geographical Review. 45:52-65, 1955. (H567)

_____. "Map makers are human: Comments on the subjective in maps." Geographical Review. 32:527-544, 1942. (H568)

_____. "Miss Semple's 'Influences of geographic environment'; notes toward a bibliography." Geographical Review. 52:346-361, 1962. (H569)

_____. "Terrae incognitae: The place of imagination in geography." Annals. 37:1-15, 1947. (H570)

_____. "The history of geography: a point of view." Annals. 26:35-37, 1936. (H571)

_____. "Training for research in political geography." Annals. 34:190-201, 1944. (H572)

Zakrzewska, B. "Trends and methods in land form geography." Annals. 57:129-165, 1967. (H573)

Zipf, G. K. "The hypothesis of the minimum equation as a unifying social principle." American Sociological Review. 22:627-650, 1942. (H574)

Zobler, L. "Decision making in regional construction." Annals. 48:140-148, 1958. (H575)

6. A number of bibliographies list the map production of foreign countries. The more important of the national bibliographies containing information on map production are the following:

Australian national bibliography, 1961- . Canberra National Library of Australia, 1961- . Monthly. (H576)

Oesterreichische Bibliographie: Verzeichnis der osterreichischen Neuerscheinungen. Bearb. von der osterreichischen Nationalbibliothek. Wien, 1946- . v. 2- . Semimonthly. (H577)

Bibliographie de belgique, 1. partie: Liste mensuelle des publications belges ou relatives à la Belgique, acquises par la Bibliothèque Royale. v. 10- . annee, 1875- . Bruxelles, Bibliothèque Royale, 1875- . v. 1- . Monthly. (H578)

Rio de Janeiro. Biblioteca Nacional. Boletim bibliografico. Rio de Janeiro, 1951- . Semiannual. (H579)

Bulgarski knigopis: mesechen bibliografski biuletin za depoziranite v Instituta Knigi i Novi Periodichni Izdanina. Sofia, 1897- . Monthly. In Cyrillic. (H580)

Canadiana, 1950- . Ottawa, National Library of Canada,

1951- . Monthly, with annual cumulations. (H581)

Det danske Bogmarked (København, Den danske Forlaeg-
gerforening). Weekly, with annual cumulations. (H582)

Suemen Kirjakauppalehti. Finsk bokhandelstidning. (Hel-
sinki, Suemen Kustannusyhdistys ja Kirjakauppias
liitte). 1907- . Semiannual. (H583)

Bibliographie de la France: journal general de l'imprimerie et de la librairie. Paris, Cercle de la Li-
brarie, 1811- . v. 1- . Weekly. (H584)

Deutsche Nationalbibliographie und bibliographie des im
Ausland erschienenen deutschsprachigen Schrifttums.
Reihe A. Reihe B. Leipzig, Verland fur Buch--und
Bibliothekswesen, 1931- . Issued in two parts; Reihe
A, Neuerscheinungen des Buchhandela, weekly, in two
parts; Reihe B, Neuerscheinungen ausserhalb des Buch-
handels. Semimonthly. (H585)

Deutsche Bibliographie: wochentliches Verzeichnis.
Frankfurt a.M., Buchhandler-Vereinigung GmbH.
1947- . Weekly. (H586)

British National Bibliography. 1950- . London, Council
of the British National Bibliography, British Museum.
1959- . Weekly, with quarterly cumulations and annual volume. (H587)

Magyar nemzeti bibliografia: bibliographia hungarica.
Kiadja ag Orszagos Szechenyi Konyvtar. 1 fuzet,
januarmarcius 1948- . Budapest, 1946- . Semimonthly. (H588)

Bibliografia nazionale italiana: nuova serie del bollettino
delle pubbicazioni italiane recevute per diritto di
stampa. Gennaio, 1958- . Firenze, 1958- . Anno
1- . Monthly. (H589)

New Zealand National Bibliography. (Wellington) National
Library of New Zealand, February 1967- . Monthly.
(H590)

Norsk Bokhandlertidende. vol. 1- . Oslo, Grendahl,
1880- . Weekly with annual cumulations. (H591)

Przewodnik Bibliograficzny: Urzedowy Wykaz Drukow
Wydanych w Rzeczypospolitej Polskiej ... B. 2 (14),
nr. 1/3- . Warszawa, Biblioteka Narodowa, 1946- .
Weekly. (H592)

SANB: Suid-Afrikaanse Nasionale Bibliografie. South
African National Bibliography, 1959- . Pretoria,
State Library 1960- . Quarterly, with annual cumula-
tions. (H593)

Boletin del deposito legal de obras impresas. Madrid
Direccion General de Archivos y Bibliotecas, 1958- .
Monthly. (H594)

El libro espanol: revista mensual ... t. 1, num. 1.
Enero, 1958- . Madrid, Inst. Nacional del Libro
Espanol, 1958- . Issued in 2 parts. Part 1, monthly;
Part 2, semimonthly. (H595)

Svenska Bokforlaggareforeningens och Svenska Bokhand-
lareforeningens Officiella Organ. Stockholm, Svenska
Bokhandel, 1952- . Weekly, with monthly, quarterly,
semiannual and annual cumulations. (H596)

Das schweizer Buch: Bibliographisches Bulletin der
schweizerischen Landesbibliothek. Le livre suisse ...
Il libro suizzero. vol. 1- ; 11 Marz 1901- . Bern
Bumplis, Benteli, 1901- . vol. 1- . (H597)

Turkiye Bibliyografyasi ... 1934- . Istanbul, Milli
Egitim Basimevi, 1935- . Quarterly. (H598)

Bibliografija Jugoslavije: Knjige, Brosure i Muzikalije,
Jan 1950- . Beograd, Bibliografski Inst. FNRJ,
1950- . Semimonthly. (H599)

7. The acquisition, cataloging, care and repair of maps as
well as the organization of map libraries is not within the
scope of this manual, but information on these matters is
found in the Stephenson and Ristow sources listed in this section.
      Most maps published by the United States government
are readily available. For information on the published
maps of the United States government, contact the Map Information Office, United States Geological Survey, Washington 25, D.C. This agency acts as a clearinghouse for map
information; index maps include the following, which are free

on request: 1) Status of topographic mapping in the United
States; 2) Status of geologic mapping in the United States; 3)
Status of horizontal control in the United States; 4) Status of
vertical control in the United States; 5) Index to topographic
maps of the United States (1:250, 000); 6) Index to topographic
and planimetric maps of each state and Puerto Rico; and concerning aerial photographs; 7) Status of aerial photography in
the United States; and 8) Status of aerial mosaics in the United
States.

The United States Geological Survey maintains several
series of topographic quadrangles available to the public for
a small fee. These include the quadrangles published at
1:24, 000 or 1:31, 680, 1:62, 500 and 1:125, 000, some of which
are national park maps, the International Map of the World,
and metropolitan area topographic maps.

Every state maintains a mapping agency, often found
under various government departments. County and city agencies or departments usually have maps of their local areas
available as well.

SUMMARY AND NOTES FOR FURTHER RESEARCH

The objective of this manual has been to introduce
some basic United States source materials useful for compiling subject bibliographies in geography. Several different
categories of source materials have been presented.

This manual has been designed to present major,
comprehensive bibliographical works first, secondary or less
comprehensive works next and, finally, specialized bibliographies in geography. Included among the comprehensive works
are Printed Library Catalogs (Chapter II) and Guides to Serials (Chapter III), while secondary sources are found in Government Publications (Chapter IV), Statistics Sources (Chapter V), and Theses and Dissertations (Chapter VI). The final chapter contains subject bibliographies in geography and
a short listing of selected readings in the field of geography.

Use of the sources contained in this manual may be
considered the geographer's first step in compiling a subject
bibliography on a chosen research problem. This manual is
highly selective and designed to emphasize contemporary
United States publications and English-language works. The
geographer with specialized research interest should now
take the next step, to compile a personal list of subject bibliographies to investigate in the course of research and then
to establish a regular routine for consulting guides to new
works in his field.

This second step in bibliography search and compilation need not be as complex as that necessary for supple-
menting comprehensive reference sources such as Winchell's
Guide.[1] Only a few sources should be consulted, the number depending upon the nature and extent of the compiler's
research interest.

Undoubtedly the most valuable source of new additions
to the geographer's current research subject bibliography is:

American Geographical Society. Library. Current geographical publications; additions to the Research

catalogue of the American Geographical Society. New
York, American Geographical Society, 1938- . 10 issues yearly. (B3)

Current geographical publications fulfills this vital
research need for a number of reasons.[2] Among these are
that Current geographical publications 1) is issued monthly
except July and August, which makes it timely; 2) is designed
to encompass all aspects of geography and selected facets of
related disciplines; 3) contains a large number of descriptive
annotations; and 4) is conveniently divided into three main
sections--General, Regional and Maps--with each section following a systematic and regional classification.

Current geographical publications is the acquisitions
list of the library of the American Geographical Society, but
it differs from ordinary library accessions lists in that it
represents the holdings of a geography-oriented library ac-
tively acquiring new publications and in that it attempts to
list every recent article, book or monograph the Librarian
deems useful to geographers. The only major category of
geography materials not included in Current geographical
publications is textbooks for the lower grades or books written in an elementary style for younger readers. The foregoing qualities make Current geographical publications the
premier source of information for new entries in an estab-
lished subject bibliography for geography research.

The primary quality required of the geography researcher interested in compiling a subject bibliography is
perseverance. Perseverance is especially necessary for the
geography bibliographer in today's world, where the search
for answers to geography research questions has been complicated by increases in the amount of information available,
the increasing rate of information generation, and the pro-
liferation of media presenting information. Only through
dedicated and intelligent use can bibliography become an effective tool and problem-solving device for geographers living in an age of information abundance.

Notes

1. For a description of this procedure, see Johnson, Olive
A., "How to write a guide to reference books." The Library Association Record, 394-397, October, 1954. (I1)

2. This analysis is based on "Current Geographical Publications" (July 1970), an unpublished manuscript by Lynn S. Mullins, Librarian of the American Geographical Society. (12)

AUTHOR INDEX

This index contains entries listed by last name of author or issuing agency. The entries for books contain citations with author, title and identification number, but the entries for serial articles contain citations with author and identification number only. The identification numbers refer to the sequence in which the works are presented in this manual.

Abrams, Charles. Man's struggle for shelter in an urbanizing world. H193
Ackerman, Edward A. Geography as a fundamental research discipline. H316
_____. H331, H332
_____. H333
_____. H334
African Bibliographic Center. A current view of Africana: a select and annotated bibliographic publishing guide. H65
Akademiia Nauk USSR. Institut Nauchnoi Informatsii. Referativng; zhurnal: geografiia. H143
Alexander, John W. Economic geography. H209
_____. H335
Alexanderson, Gunnar, and Goran, Norstrom. World shipping: an economic geography of ports and seaborne trade. H237
Alnwick, Herbert. A geography of commodities. H238
America en cifras. F17
American Association of Petroleum Geologists. Committee on Preparation of Projection Slides. AAPG slide manual: a guide to the preparation and use of projection slides. H312
American Bibliographical Center. America: history and life, a guide to periodical literature. H52
American Chemical Society. Chemical abstracts. H136
American Geographical Society. Library. Current geographical publications: additions to the Research catalogue of the American Geographical Society. B3
_____. Research catalogue. B2
_____. Map Department. Index to maps in books and periodicals. B65

137

_____. Readings in geography of North America; a selection of articles from the Geographical Review. H262

American Geological Institute. Committee on Education. Outstanding aerial photographs in North America. H313

American Geophysical Union. Bibliography of hydrology, United States of America. H17

American Health Association. Committee on the Hygiene of Housing. An appraisal method for measuring the quality of housing. H194

American Meteorological Society. Meteorological and geoastrophysical abstracts. B37

American Society of Photogrammetry. Manual of photographic interpretation. H106

American Society of Photogrammetry. Manual of photographic interpretation. H307

American Universities Field Staff. A select bibliography: Asia, Africa, Eastern Europe, Latin America. B57

American Universities Field Staff. H53

_____. H54

American Water Resources Association. Hydata, 1965- . H18

Ames, John G. Comprehensive Index to the publications of the United States government, 1881-1893. E18

Anderson, Marc. A working bibliography of mathematical geography. B24

Anderzhon, Mamie L. A selected bibliography on geography education for curriculum committees. H111

Andriot, John L. Guide to United States government serials and periodicals. B21

_____. Guide to United States government statistics. B28

Annotated bibliography of economic geology. H128

Appleton, J. H. A morphological approach to the geography of transport. H223

Applied science and technology index. D12

Arab States Fundamental Education Centre. Social Science Division. Statistical sources of the Arab states; a comprehensive list. F21

Arctic Institute of North America. Arctic bibliography. H73, H74

Askling, John. D16, D17

Association de Geographes Français. Bibliographie geographique internationale. H145

Aufricht, Hans. Guide to League of Nations publications: a bibliographical survey of the work of the League, 1920-1947. E10

Australia. National Library. Australian national bibliography. H69

Australian national bibliography, 1961- . H463

Avicenne, Paul. Bibliographical services throughout the world, 1960-1964. A17
Ayer, N. W., and Son (firm). N. W. Ayer and Son's Directory of newspapers and periodicals. D1

Bachi, R. H336
Baird, Patrick D. The Polar World. H297
Baker, O. E. H337
Ballabon, M. B. H338
Barnum, H. G., Kasperson, R., and Kiuchi, S. Central place studies ... Supplement, 1965. B26
Barrows, H. H. H339
Bateman, Alan M. (ed.) Economic geology. H234
Baum, W. A. and Court, A. H340
Beaujeu-Garnier, J. Geography of population. H199
Becker, Joseph and Hayes, Robert M. Information storage and retrieval. A2
Beckmann, M. J. H341
Bengtson, Nels A., and Van Royen, William. Fundamentals of economic geography; an introduction to the study of resources. H210
Bennett, A. S. H342
Bercaw, Louise O. Bibliography on land utilization 1918-1936. H39
Berry, B. J. L. H343
_____. H344
_____. H345
_____, and Garrison, W. L. H346
_____, and Hankins, Thomas D. A bibliographic guide to the economic regions of the United States. B58
_____, and Pred, Allan R. Central place studies: a bibliography of theory and applications. B25
Bestermann, Theodore. A world bibliography of bibliographies and of bibliographical catalogues, calendars, abstracts, digests, indexes and the like. A12
Bestor, George C., and Jones, Holway R. City planning: a basic bibliography of sources and trends. B43
Bibliografia nazionale italiana: nuova serie del bollettino delle pubblicazioni italiane recevute per diritto di stampa. H589
Bibliografija Jugoslavije: Knjige, Brosure i Muzikalije. H599
Bibliographic index: a cumulative bibliography of bibliographies. A13
Bibliographic cartographique internationale, 1936- . H85
Bibliographie de belgique, 1. partie: Liste mensuelle des publications belges ou relatives à la Belgique, acquises

par la Bibliothèque Royale. H578
Bibliographie de la France: journal general de l'imprimerie
et de la librairie. H584
Bibliography of soil science, fertilizers and general agron-
omy, 1959-1962. B41
Biological abstracts: reporting the world's biological re-
search literature. D13
Birch, J. W. H347
Bishop, William Warner. C15
Black, Dorothy M. (comp.) Guide to lists of master's the-
ses. G6
Blaisdell, Ruth F. et. al. Sources of information in trans-
portation. H34
Blake, Sidney F. Geographical guide to floras of the world.
B39
Blanchard, J. R. and Ostvold, Harold. The literature of
agricultural research. H30
Blaut, J. M. H348
_____. H349
Blumenfeld, H. H350
Body, Alexander C. Annotated bibliography of bibliographies
on selected government publications and supplementary
guides to the Superintendent of Documents classification
system. E16
Boesch, Hans H. A geography of world economy. H211
Boggs, Samuel W., and Lewis, Dorothy C. The classifica-
tion and cataloging of maps and atlases. H302
Bogue, Donald J. and Beale, Clavin L. Economic areas of
the United States. F114
Boletin del deposito legal de obras impresas. H594
Books in print; an author-title index to the Publishers' trade
list annual. C4
Borchert, J. R. H351
Bourne, C. P. Methods of information handling. A18
Bowker, R. R. A provisional list of the official publications
of the several states of the United States from their or-
ganizations. E37, E47
Bowman, Isiah. Geography in relation to the social sciences.
H317
_____. H352, H353
_____. H354
Boyce, Ronald R. H340
Boyd, Anne M. United States government publications. E13
BPR: American book publishing record. C7
Branch, Melville C. Comprehensive urban planning: a se-
lective annotated bibliography with related materials. B44
Brigham, A. P. H355

Brimmer, Brenda (and others). A guide to the use of United
Nations documents. E11
British Museum. Department of Printed Books. Map Room.
Catalogue of printed maps, plans, and charts in the Bri-
tish Museum. H75
British National Bibliography. 1950- . H587
Broek, Jan O. M. Geography: its scope and spirit. H318
_____. H356
_____. H357
_____, and Innis. H358
Brookfield, H. C. H359
Brown, Everett S. Manual of government publications:
United States and foreign. E1
Brown, Lloyd A. The story of maps. H303
Brown, Ralph H. Historical geography of the United States.
H255
_____. H360
Brown, W. J. H361
Browning, Clyde E. A bibliography of dissertations in geog-
raphy: 1901 to 1969. B30
Brunhes, Jean. Human geography: an attempt at a positive
classification, principles and examples. H319
Bryan, K. H362
Buchanan, R. O. H363
Buckman, Harry O., and Brady, Nyle C. The nature and
property of soils. H184
Bulgarski knigopis: mesechen bibliografski biuletin za
depoziranite y Instituta Knigi i Novi Periodichni Izdanina.
H580
Bunge, Charles A. A22
Bunge, William W. Theoretical geography. H320
_____. H364
_____. H365
Bunting, Brian T. The geography of soils. H185
Burdette, Franklin L. (and others). Political science: a se-
lected bibliography of books in print with annotations. H49
Burker, L. V. H366
Burton, I. H367
Butzer, Karl W. Environment and archeology: an introduc-
tion to Pleistocene geography. H256

Calef, W. C. H368
California. University. Berkeley. Bancroft Library. In-
dex to printed maps. H76
_____. _____. University at Los Angeles. Committee on
Latin American Studies. Statistical abstract of Latin
America, 1955- . F16

Canadiana, 1950- . Ottawa, National Library of Canada, 1951- . H581
Carlson, Albert S. Economic geography of industrial materials. H235
Carnegie Institution. Index of economic material in documents of the United States. Comp. by Adelaide Hasse. E38
Carol, H. H369
Carrothers, G. P. H370
Carson, Rachel L. The sea around us. H175
Carter, H. C. H371
Centre for Urban Studies. Land use planning and the social sciences: a selected bibliography. H40
Centre National de la Recherche Scientifique. Repertoire des principaux périodiques d'interet géographique cites dans la Bibliographie Géographique Internationale. H153
Chapin, F. Stuart, Jr. Selected references on urban planning, methods and techniques. H23
_____. Urban Land Use Planning. H188
Childe, V. Gordon. Man makes himself. H257
Childs, James B. Government document bibliography in the United States and elsewhere. E2
Chisholm, M. H372
Christian Science Monitor. Subject index of the Christian Science Monitor, vol. 1- . D15
Chorley, R. J. H373
Chronic, J., and Chronic, H. Bibliography of theses in geology. H131
_____. Bibliography of theses in geology. H132
Clapp, V. W. D18
Clark, A. H. H374
Clarke, J. S. (and others). H119
Clarke, John I. Population geography. H200
Clarks, Edith E. E48
Claussen, Martin P. and Friis, Herman R. Descriptive catalog of maps published by Congress, 1817-1843. H77
Cochran, W. F.; Mosteller, F., and Tukey, J. W. H375
Cohen, Saul B. Geography and politics in a world divided. H245
_____. H352
Colby, Charles C. H376
_____. H377
_____. H378
_____, and Foster, Alice. Economic geography; industries and resources of the commercial world. H236
Cole, Arthur H. Measures of business change; a Baker Library index. F115

Cole, Dorothy E. A29
Cole, John P., and German, Frank. A geography of the
   U.S.S.R.; the background to a planned economy. H280
Collison, Robert L. Bibliographies, subject and national; a
   guide to their contents, arrangement and use. A15
_____. Bibliographical services throughout the world, 1950-
   1959. A16
Conover, Helen F. (comp.). Introduction to Europe: a se-
   lective guide to background reading. H62
Coulter, Edith M., "Selected List of References on State
   Documents," in Wilcox, Jerome K., (ed.), Manual on the
   use of state publications. E46
Council of State Governments. Book of the states, vol. 1- .
   E26
_____. Index to council of state government publications.
   E27
Courtney, William P. A register of national bibliography;
   with a selection of the chief bibliographical books and ar-
   ticles printed in other countries. C12
Cressey, George B. Asia's land and peoples; a geography
   of one-third of the earth and two-thirds of its people.
   H287
_____. Soviet potentials: a geographic appraisal. H281
Critchfield, Howard J. General climatology. H163
Crowe, P. R. H379
Culver, Dorothy C. Land utilization: a bibliography. H41
Cumberland, Kenneth B. Southwest Pacific; a geography of
   Australia, New Zealand and their Pacific island neighbors.
   H292
_____, and Fox, James W. New Zealand a regional
   view. H293
Cumulative book index: a world list of books in the English
   language. C8
Curry, L. H380

Daggett, Stuart. Principles of inland transportation. H224
Dale, Doris C. The manual retrieval of government publica-
   tions. B70
Dansereau, Pierre M. Biogeography: an ecological per-
   spective. H180
Darlington, Philip J. Zoogeography: the geographical dis-
   tribution of animals. H181
Darton, N. H. H113
Das schweizer Buch: Bibliographisches Bulletin der schwei-
   zerischen Landesbibliothek. H597
David, Charles W. C16
Davies, J. L. H381

Davis, W. M. H382, H383, H384, H385, H386
De Blij, Harm J. Systematic political geography. H246
De Geer, S. H387
Denman, Donald R. Bibliography of rural land economy and
    land ownership, 1900-1957. H42
Det danske Bogmarked. H582
Deutsche Bibliographie: wochentliches Verzeichnis. H586
Deutsche Nationalbibliographie und bibliographie des im Aus-
    land erschienenen deutschsprachigen Schrifttums. H585
Dewey, Melvil. Dewey decimal classification and relative
    index. A20
Dewhurst, J. Fredric; Coppock, John O.; Yates, P. Lamar-
    tine, and Associates. Europe's needs and resources;
    trends and prospects in eighteen countries. H273
Dickinson, Robert E. The West European city. H274
_____, and Howarth, Osbert J. R. The making of geography.
    H321
Dissertation abstracts international: abstracts of disserta-
    tions and monographs in microform. B31
Dissertations of the sixties. G4
Doctoral dissertations accepted by American universities,
    1933/34-1954/55. G2
Dodge, S. D. H388
Donahue, Roy L. Soils: an introduction to soils and plant
    growth. H186
Dryer, C. R. H389
Dunn, E. S. H390

East, William Gordon. An historical geography of Europe.
    H258
_____, and Spate, Oscar H. K. (eds.). The changing map
    of Asia. H288
Editor and Publisher. International yearbook number for
    1920- . D2
Edward Stanford, Ltd. International map bulletin. H86
Edwards, Richard A. Index digest of state constitutions,
    2nd ed. E28
Egler, F. E. H391
El libro espanol: revista mensual. H595
Eldridge, Hope T. The materials of demography: a se-
    lected and annotated bibliography. H25
Embree, John F. Books on Southeast Asia: a select bib-
    liography. H67
Engineering Index, Inc. Engineering index. H137
Estall, R. C., and Buchanan, Ogilvie R. Industrial activity
    and economic geography. H216
Ettinger, Jan. Towards a habitable world; tasks, problems
    and methods, acceleration. H195

European Cultural Centre. The European bibliography, ed.
by Hjalmar Pehrsson and Hanna Wulf. H62
Eyre, Samuel R. Vegetation and soils, a world picture.
H179

Faison, Georgia H. E49
Felland, Nordis. United Nations publications useful to geog-
raphers. B20
Fenneman, Nevin M. Physiography of the Eastern United
States. H156
_____. Physiography of the Western United States. H157
_____. H392
_____. H393
Field, Henry H. Bibliography on Southwestern Asia. H68
Finch, V. C. H394
_____. H395
Fischer, William A. Remote sensing research in the United
States: a summary. H107
Fisher, C. A. H396
Fitzgerald, W. H397
Fleure, H. J. H398
Florida. University and United States. Library of Congress.
Hispanic Foundation. Handbook of Latin American Studies.
H60
Floyd, B. N. H399, H400, H401
Food and Agriculture Organization of the United Nations.
Bibliography on land tenure. B51
_____. _____. H43
Forgotson, J. M. H402
Forthcoming books. C9
Freeman, T. W. A hundred years of geography. H322

Gabler, Robert E. (ed.). A handbook for geography teach-
ers. B68
_____. _____. H314
Garrison, W. L. H403
_____, and Marble, D. F. H404
_____, (and others). Studies of highway development and
geographic change. H225
Gauld, W. A. H405
Geographical Abstracts. H144
Geographical Digest. H6
Geographisches Jahrbuch. VEB Hermann Haack, Gotha.
H147
Geological Society of America. Bibliography and index of
geology. H127
_____. Bibliography and index of geology exclusive of North
America. H126

_____. Bibliography of military geology and geography. H51
Geoscience Information Society. Geologic field trip guide
    books of North America. H135
German Cartographical Society and the Bundeanastalt fur
    Landeskunde und Raumforschung. Bibliotheca cartographi-
    ca. H87
Gilbert, E. W. H406
Gill, Richard T. Economic development; past and present.
    H240
Ginsburg, Norton S. (and others). The pattern of Asia.
    H289
Gottmann, Jean. A geography of Europe. H275
_____. H407
_____. Megalopolis: the urbanized north-eastern seaboard
    of the United States. H189
Gould, Laurence M. The Polar regions in the relation to
    human affairs. H298
Gould, P. R. H408
Gourde, Leo. C17
Government Affairs Foundation, Inc. Metropolitan communi-
    ties; a bibliography, with special emphasis upon govern-
    ment and politics. E42
_____. E43
"Government Publications." E50
"Government publications and their use." E51
Graves, William B. (and others). American state govern-
    ment and administration, a state-by-state bibliography of
    significant general and special works. E30
Greenhood, David. Mapping. H304
Gregor, Howard F. Environment and economic life. H212
Gregory, Winifred (ed.). List of the serial publications of
    foreign governments, 1815-1931. E3
Gropp, Arthur E. (comp.). A bibliography of Latin Ameri-
    can bibliographies. B59
Grotewald, A. H409
Gunther, Edgar, and Goldstein, Frederick A. Current
    sources of marketing information; a bibliography of pri-
    mary marketing data. H1
Gywer, Joseph A., and Waldron, Vincent G. Photo inter-
    pretation techniques; a bibliography. B66

Hagood, M. J. H410
Hailey, William M. H. (Baron). An African survey; a study
    of the problems arising in Africa south of the Sahara.
    H282
Hall, B. H411
_____, and Noh, Toshio. Japanese geography: a guide to
    Japanese reference and research materials. B62

Hance, William A. Geography of modern Africa. H283
Hannah, William. H99
Hanson, R. M. H412
Hardin, Ruth. E52
Hare, F. Kenneth. The restless atmosphere. H164
Harman, Marian. C18
Harmon, Robert B. Political science: a bibliographical
    guide to the literature. B53
Harris, B. F. D. H413
Harris, Chauncy D. Annotated world list of selected cur-
    rent geographical serials in English: including an appendix
    of major serials in other languages with regular supple-
    mentary or partial basic use of English. B14
    _____. Bibliographies and reference works for research in
    geography. B32
    _____, and Fellmann, Jerome D. International list of geo-
    graphical serials. B15
Harris, K. G. A30
Harris, William J. Guide to New Zealand reference material
    and other sources of information. H70
Hart, J. F. H414
    _____. H415
Hartshorne, Richard. Perspective on the nature of geography.
    H323
    _____. The nature of geography: a critical survey of cur-
    rent thought in the light of the past. H324
    _____. H416, H417, H418, H419, H420
    _____. H421
    _____. H422
    _____, and Dicken, S. M. H423
Hartsman, G. W. H424
Harris, C. D. H425
Harvey, Anthony P. H129
    _____. _____. H141
Harvey, David. Explanation in Geography. H325
    _____. H426
Hatherton, Trevor (ed.). Antarctica. H299
Hauser, Philip M., and Duncan, Otis (eds.). The study of
    population; an inventory and appraisal. H201
    _____, and Leonard, William R. (eds.). Government sta-
    tistics for business use. H2
    _____, and Schnore, Leo F. (eds.). The study of ur-
    banization. H190
Hazlewood, Arthur (comp.). The economics of "underdevel-
    oped" areas: an annotated reading list of books, articles
    and official publications. H45
Helburn, N. H42

Heller, C. F. H428
Heller, Robert L. (ed.). Geology and Earth Sciences source-
    book for elementary and secondary schools. H315
Herbertson, A. J. H429
Hewes, Leslie. H10, H11, H12, H430
Higbee, Edward C. American agriculture; geography, re-
    sources, conservation. H218
Higgins, Benjamin H. Economic development; principles,
    problems, and policies. H241
Hill, Roscoe R. The National archives of Latin America.
    E8
Hilton, Ronald. Handbook of hispanic source materials and
    research organizations in the United States. H61
Hirshberg, Herbert S. and Melinat, Carl H. Subject guide
    to United States government publications. E14
Hodgson, James G. Official publications of American coun-
    ties, a union list. E41
Hoffman, George W. (ed.). A geography of Europe, includ-
    ing Asiatic U. S. S. R. H276
Holleb, Doris B. Social and economic information for urban
    planning. B45
Hollingsworth, Josephine B. E53
Horecky, Paul L. (ed.). Basic Russian publications, an an-
    notated bibliography on Russia and the Soviet Union. H63
Hudson, G. D. H431
_____. H432
Huff, W. H. D19
Huntington, E. H433

Index to American doctoral dissertations, 1955/56- . G3
Index to theses accepted for higher degrees in the universities
    of Great Britain and Ireland. G12
"Indexing of book reviews in periodicals. " D20
Information please almanac, atlas and yearbook, 1947- .
    F121
Institut fur Landeskunde. Abteilung dokumentation. H150
_____. Documentatio Geographica. Geographische zeit -
    schriften-und serien-literatur. H149
_____. Verzeichnis der geographischen zeitschriften, period-
    ischen veroffentlichengen und schriftenreihen Deutschlands
    und der in den letzteren erschienenen arbeiten. H155
Institut International de Statistique. Office Permanent. An-
    nuaire international de statistique. F4
_____. F3
Interagency Committee on Oceanography. Bibliography of
    oceanographic publications. H19
Inter-American Statistical Institute. Bibliography of selected
    statistical sources of the American nations. F14

_____. Monthly list of publications received. H3
International Geographical Union. Commission on Coastal
    Sedimentation. H14
_____. Special Commission on the Humid Tropics. B56
International Hydrographic Bureau. International Hydro-
    graphic Bulletin. H88
Isard, W. H434

Jackson, Barbara (Ward). The rich nations and the poor
    nations. H242
Jackson, Benjamin D. Guide to the literature of botany.
    H21
Jackson, W. A. Douglas, (ed.). Politics and geographic
    relationships: readings on the nature of political geogra-
    phy. H247
Jacobstein, J. M. D21
James, Preston E. Latin America. H268
_____. H435, H436, H437
_____, and Jones, Clarence F. (eds.). American geography:
    inventory and prospect. H326
Jefferson, M. H438
_____. H439
Johnson, D. H440
Johnson, L. J. H441
Johnson, Olive A. I1
Jones, E. H442
Jones, S. B. H443
_____. H444
Jones, W. D. and Finch, V. C. H445
_____, and Sauer, C. O. H446
Jorre, Georges. The Soviet Union, the land and its people.
    H279
Journal of Geography. H89

Kansky, Karol J. Structure of transportation networks; re-
    lationships between network geometry and regional char-
    acteristics. H226
Kaplan, Louis. A31
Katz, Saul M. and McGowan, Frank. A selected list of
    readings on development. H46
Katzman, Raphael. Modern hydrology. H171
Kendrew, Wilfred G. The climates of the continents. H165
Kent, Allen. Specialized information centers. A1
Kerker, Ann E., and Schlundt, Esther M. Literature
    sources in the biological sciences. H22
Kesseli, J. E. H447
Kimble, George H. T. Tropical Africa. H284

Kimble, George H. T., and Good, Dorothy (eds.). Geography of the northlands. H300

Kindleberger, Charles P. Economic development. H243

King, Cuchlaine A. M. An introduction to oceanography. H176

King, L. J. H448

_____. H449

King, Phillip B. The evolution of North America. H158

King, R. R. (and others). H117

Kinney, Mary R. Bibliographical style manuals: a guide to their use in documentation and research. A25

Kirk, W. H450

Klages, Karl H. W. Ecological crop geography. H219

Klimm, L. E. H451

Koeppe, Clarence E., and DeLong, George C. Weather and climate. H166

Kohn, C. F. H452

_____. H453

Kollmorgen, W. M. H454

Küchler, A. W. H455

_____. H456

Kuenen, Philip H. Realms of water; some aspects of its cycle and nature. H172

Lancaster, Henry O. Bibliography of statistical bibliographies. B27

Lander der Erde; politisch-ökonomisches Handbuch. F11

Landsberg, Helmut E. Physical climatology. H167

Larsen, Knud. National bibliographical services, their creation and operation ... Paris, UNESCO, 1953. C13

Lauche, Rudolf (ed.). World bibliography of agricultural bibliographies. H31

LaValle, P., McConnell, H. and Brown, R. G. H457

League of Nations. Statistical yearbook of the League of Nations; 1926-1942/44. F23

Leidy, W. Phillip. A popular guide to government publications. E15

Leighly, J. H458, H459

Lewis, P. W. H460

Lewthwaite, Gordon R. H461

_____, Price, Edward T., Winters, Harold A. A geographical bibliography for American college libraries; a revision of A basic geographical library, a selected and annotated book list for American colleges. B33

Licklider, J. C. R. Libraries of the future. A4

Lively, C. E. and Gregory, C. C. H462

Lloyd, Gwendolyn. E33

Lobeck, Armin K.   Geomorphology.   H159
Logan, Marguerite.   Geographical bibliography for all the
   major nations of the world; selected books and magazine
   articles.   H55
Long, H. K.   H130
Lord, Russell.   The care of the earth.   H229
Lösch, A.   H463
Lowenthal, D.   H464, H465
Lucien, F.   H466
Lueder, Donald R.   Aerial photographic interpretation;
   principles and application.   H308
Lukermann, F.   H467, H468, H469
Lystad, Robert A.   The African world; a survey of social
   research.   H285

MacFadden, C. H.   H470
Mackay, J. R.   H471
Mackinder, Halford J.   Britain and the British seas.   H251
_____.   H472
Magyar nemzeti bibliografia: bibliographia hungarica.   H588
Mahan, Alfred T.   The influence of sea power upon history
   1660-1783.   H252
Maichel, Karol.   Guide to Russian reference books.   H64
Malin, J. C.   H473
"Manual of government publications, United States and for-
   eign."   E54
Marschner, F. J.   Land use and its patterns in the United
   States.   H230
Marshall, Mary J.   Union list of higher degree theses in
   Australian university libraries.   G9
Martinson, Tom L.   A24
Massachusetts Oceanographic Institutions (Woods Hole).   A
   partial bibliography of the Indian Ocean.   H20
Masters abstract: abstracts of selected masters theses on
   microfilm.   G5
Mattice, W. A.   H474
May, Jacques M.   Studies in disease ecology.   H205
_____.   H475
Mayer, H. M.   H476
_____.   H477
_____, and Kohn, Clyde F. (eds.).   Readings in urban geog-
   raphy.   H191
McCameron, Fritz.   FORTRAN logic and programming.   B69
McCarty, H. H.   H478
_____.   H479
McColvin, L. R.   D22

152 Library Research in Geography

McCrum, Blanche P., and Jones, Helen D. Bibliographical procedures and style: a manual for bibliographers in the Library of Congress. A26
McManis, Douglas R. Historical geography of the United States: a bibliography--excluding Alaska and Hawaii. B55
McMurry, K. C. H480
McNee, R. B. H481
Mead, William R., and Brown, E. H. The United States and Canada; a regional geography. H267
Meigs, P. H482
Meinzer, Oscar E. (ed.). Hydrology. H173
Mellor, Roy E. H. Geography of the U.S.S.R. H278
Metcalf, Kenneth H. Transportation information sources; an annotated guide to publications, agencies and other data sources concerning air, rail, water, road and pipeline transportation. H35
Military Engineer. H90
Miller, G. J. H408
Miller, Marvin A. C19
Miller, R. Willard. A geography of manufacturing. H217
Miller, Victor C. Photogeology. H309
Mineralogical Society of Great Britain and Mineralogical Society of America. Mineralogical abstracts. H139
Monkhouse, F. J., and Wilkinson, H. R. Maps and diagrams; their compilation and construction. H305
Morgan, W. B. and Moss, R. P. H483
Mueller, Bernard. A statistical handbook of the North Atlantic area. F10
Mullins, Lynn S. I2
Mumford, Lewis. From the ground up. H196
_____. The city in history: its origins, its transformations, and its prospects. H192
Municipal year book, 1934- ; the authoritative resumé of activities and statistical data of American cities. E44
Murphey, R. E. H409
Murphy, R. C. H485

Nanson, Fridtjof. In northern mists; Arctic exploration in early times. H301
NASA. Earth Resources Research Data Facility Index. H108
_____. Thermal Infrared Imagery in Urban Studies. H109
National Academy of Sciences. National Research Council. Earth Sciences Division. The Science of geography: report of the ad hoc committee on geography. H327
National Association of Home Builders of the United States. Basic texts and reference books on housing and construction: a selected, annotated bibliography. B46

National Association of Housing and Redevelopment Officials.
    Summary of housing year: bibliography of housing litera-
    ture. H24
National Association of Manufacturers of the United States of
    America. Bibliography of economic and social study ma-
    terial issued by the National Association of Manufacturers.
    H27
National Association of State Libraries. Public Documents
    Clearing House Committee. Checklist of legislative jour-
    nals of the states of the United States of America. E34
National Industrial Conference Board. Economic almanac for
    1940- . H4
National Research Council. Committee on Oceanography.
    Oceanographic information sources; a staff report. B38
New serial titles, 1950-1960; supplement to the Union list of
    serials, 3rd ed. B18
New York Times index, vol. 1- . D14
New Zealand National Bibliography. H590
Nicholson, N. L. H100
Nickles, J. M. H114, H115
Nineteenth Century readers' guide to periodical literature,
    1890-1899, with supplementary indexing, 1900-1922. D8
Norsk Bokhanlertidende. H591
Northwestern University. Library of the Transportation Cen-
    ter. Current literature in traffic and transportation. H36

Oceanic Research Institute. Oceanic index. H140
Oesterreichische Bibliographie: Verzeichnis der osterreich-
    ischen Neuerscheinungen. H577
Olson, Ralph E. The literature of regional geography; a
    checklist for university and college libraries. H56
Olsson, Gunnar. Distance and human interaction: a review
    and bibliography. B42
Ottawa. Canadian Bibliographic Centre. Canadian graduate
    theses in the humanities and social sciences, 1921-1946.
    G10
_____. National Library of Canada. Canadian theses:
    Theses Canadiennes, 1960/61- . G11
Ottoson, Howard W. (ed.). Land use policy and problems
    in the United States. H231

Palfrey, Thomas R. and Coleman, Henry E. Guide to bib-
    liographies of theses, United States and Canada. G8
Parkins, A. E. H486
Paterson, John H. North America; a regional geography.
    H263

Paullin, Charles O. (ed.). Atlas of the historical geography
of the United States. H259
Pearl, Richard M. Guide to geologic literature. H13
Peltier, Louis C. (ed.). Bibliography of military geography.
B54
_____, and Pearcy, G. Etzel. Military geography.
H253
Pelzer, Karl J. Selected bibliography on the geography of
Southeast Asia. B63
Penrose, E. F. H487
Petterssen, Sverre. Introduction to meteorology.
H168
Philbrick, A. K. H488
Platt, Robert S. Latin America; countrysides and united
regions. H269
_____. H489, H490, H491, H492
Polunin, Nicholas V. Introduction to plant geography and
some related sciences. H182
Poole, S. P. H493
Poole's Index to periodical literature, 1802-81. D7
Poore, Benjamin P. A descriptive catalog of the government
publications of the United States, September 5, 1774- .
E17
Porter, Philip W. A bibliography of statistical cartography.
B64
_____. H494
Pounds, Norman J. G. Political geography. H248
_____. Europe and the Soviet Union, 2nd ed. H277
Press, Charles and Williams, Oliver. State manuals, blue
books, and election results. E39
Princeton University. Office of Population Research and
Population Association of America. Population Index.
H26
Proctor, Cleo V., Jr. H112
Professional Geographer. H91
Przewodnik Bibliograficzny: Urzedowy Wykaz Drukow Wy-
danych w Rzeczpospolitej Polskiej. H592
Public Affairs Information Service. Bulletin. D11
Publishers' trade list annual. C3
Publishers' Weekly: the American book trade journal. C6
Pullen, William R. A check list of legislative journals is-
sued since 1937 by the states of the United States of
America. E35
Putman, D. F., and Kerr, D. P. A regional geography of
Canada. H264
Putnam, William C. Geology. H160

Quant, R. E.   H495

Raisz, Erwin J.   General cartography.   H306
_____.   H496
Rand McNally and Company.   Geographic Research Depart-
    ment.   Map library acquisitions bulletin: a list of atlases,
    maps and books received by the map library.   H92
Raup, H. M.   H497
Reader's digest almanac, 1966- .   F122
Readers' guide to periodical literature, 1900- .   D9
Real Sociedad Geografica.   Catalogo de la biblioteca.   H152
Reeds, L. G.   H498
Renner, G. T.   H499
ReQua, Eloise G., and Statham, Jane.   The developing na-
    tions: a guide to information sources concerning their eco-
    nomic, political, technical and social problems.   B52
Reynolds, R. B.   H500
Riehl, Herbert.   Introduction to the atmosphere.   H169
Rio de Janeiro.   Biblioteca Nacional.   Boletim bibliografico.
    H579
Ristow, Walter W.   H101
Robinson, A. H.   H501
_____, and Bryson, R. A.   H502
Robinson, Kathleen W.   Australia, New Zealand and the
    Southwest Pacific.   H294
Roorbach, G. B.   H503
Rosenthal, Joseph A.   E55
Royal Geographical Society.   Current geographical periodicals:
    a hand-list and subject index of current periodicals in the
    library of the Royal geographical society.   H154
_____.   New Geographical literature and maps: additions to
    the Library of the Royal Geographical Society, 1951- .
    H93
_____.   New geographical literature and maps: addi-
    tions to the library of the Royal Geographical So-
    ciety, 1951- .   H151
Russell, R. J.   H504

Sachet, Marie H. and Fosberg, Francis R.   Island bibliogra-
    phies: Micronesian botany, land environment and ecology
    of coral atolls, vegetation of tropical Pacific islands.   H71
SANB: Suid-Afrikaanse Nasionale Bibliografie.   South African
    National Bibliography, 1959- .   H593
Sauer, Carl O.   Agricultural origins and dispersals.   H220
_____.   H505
Schaefer, F. K.   H506

Schimper, Andreas F. W. Plant-geography upon a physiological basis. H183
Schmeckebier, Laurence F., and Eastin, Roy B. Government publications and their use. B22
Schmid, C. F. and MacCannell, E. H. H507
Schnore, L. F. H508
Schoy, C. H509
Schulte, O. W. H510
Schultheiss, Louis A., Culbertson, Don and Heilinger, Edward M. Advanced data processing in the university library. A3
Shannon, Lyle W. Underdeveloped areas, a book of readings and research. H244
Sheehy, Eugene P. Guide to reference works. A8
_____. A9, A10
Shores, Louis. Basic references; an introduction to materials and methods. A14
Shreve, F. H511
_____. H512
Siddall, William R. Transportation geography: a bibliography. B50
_____. H513
Smailes, A. A. A5
Smith, Harold T. U. Aerial photographs and their applications. H310
_____. H514
Smith, Harriet W. H142
Smith, Roger C. Guide to the literature of the zoological sciences. B40
Smith, J. Russell, and Phillips, M. Ogden. North America. H265
Smith, W. H515
Snyder, David E. A19
Social sciences and humanities index. D10
Société de Geographie de Paris. Acta geographica. H146
Spate, O. H. K. H516, H517, H518
Special Libraries Association. Geography and Map Division Bulletin. H94
Spencer, Joseph E. Asia, east by south, a cultural geography. H290
_____, and Horvath, R. J. H519
Spurr, Stephen H. Photogrammetry and photo-interpretation, with a section on applications to forestry. H311
Spykman, N. J. H520
Stamp, L. Dudley. Applied geography. H328
_____. Asia, a regional and economic geography. H291
_____. The geography of life and death. H206

_____. Some aspects of medical geography. H207
_____. Africa, a study in tropical development. H286
_____. H521
Statesman's year-book; statistical and historical annual of the
   states of the world, 1864- . F5
Statistical handbook of Middle Eastern countries. F22
Staton, Frances M. A bibliography of Canadiana; being items
   in the public library of Toronto, Canada, relating to early
   history and development of Canada. H57
Staveley, Ronald. Guide to unpublished research materials.
   G7
Steiner-Prag, E. F. D23
Stephenson, Richard W. H102
Stevens, Benjamin H., and Brackett, Carolyn A. Industrial
   location: a review and annotated bibliography of theoreti-
   cal, empirical and case studies. B48
Steward, Julian H. (ed.). Handbook of South American In-
   dians. H270
Stewart, Charles F. and Simmons, George B. (comps.). A
   bibliography of international business. H47
Stewart, C. T., Jr. H522
Stewart, J. Q. H523
_____, and Warntz, W. H524
Stokes, G. A. H525
Stone, K. H. H526
_____. H527
Strahler, A. N. H528
Subject guide to books in print: and index to the Publishers'
   trade list annual. C5
Suemen Kirjakauppalehti. Finsk bokhandelstidning. H583
Surveying and Mapping. H95
Svenska Bokforlaggareforeningens och Svenska Bokhanlare-
   foreningens Officiella Organ. H596
Sverdrup, Harold U., Johnson, Martin W., and Fleming,
   Richard H. The oceans: their physics, chemistry, and
   general biology. H177
Sykes, Percy. A history of exploration from the earliest
   times to the present day. H260

Taylor, Clyde R. H. A Pacific bibliography: printed material relating to the native peoples of Polynesia, Melanesia
   and Micronesia. H72
Taylor, E. G. R. H529
Taylor, T. Griffith. Australia; a study of warm environments and their effect on British settlement. H295

Taylor, Griffith (ed.). Geography in the twentieth century;
  a study of growth, fields, techniques, aims, and trends.
  H329
_____. H530
Teggert, F. J. H531
Texas, University. Population Research Center. Interna-
  tional population census bibliography. H5
Textbooks in print, including teaching materials, 1956- .
  C10
Thiele, Walter. Official map publications: a historical
  sketch, and a bibliographical handbook of current maps
  and mapping services in the United States, Canada, Latin
  America, France, Great Britain, Germany and certain
  other countries. H78
Thom, E. M. H116
Thoman, Richard S. The geography of economic activity;
  an introductory world survey. H214
_____. H532
Thomas, E. N. H533
Thomas, William L., Jr. (ed.). Man's role in changing
  the face of the earth. H261
Thompson, K. H534
Thompson, Warren S., and Lewis, David T. Population
  problems. H202
Thornbury, William D. Principles of geomorphology, New
  York, Wiley, 1954. H161
_____. Regional geomorphology of the United States.
  H162
Thornthwaite, C. W. H535
_____. H536
_____. H537
_____. H538
Todd, David K. Ground water hydrology. H174
Tompkins, Dorothy C. State government and administration;
  a bibliography. E40
Toronto. Public Library. The Canadian catalog of
  books published in Canada, about Canada, as
  well as those written by Canadians, 1921-1949.
  H58
Tower, W. S. H539
Trewartha, Glenn T. An introduction to climate.
  H170
_____. H540
Troll, C. H541
Tromp, Solco W. Medical biometeorology: weather, cli-
  mate and the living organism. H208

Turabian, Kate L. A manual for writers of term papers,
   theses, and dissertations. A27
_____. Student's Guide for writing college papers. A28
Turkiye Biblyografyasi. H598
Twentieth Century Fund. Survey of tropical Africa; select
   annotated bibliography of tropical Africa. B61

Ullman, Edward L. American commodity flow; a geographic
   interpretation of rail and water traffic based on principles
   of spatial interchange. H227
_____. H543, H544
Ulrich's international periodicals directory; a classified guide
   to current periodicals, foreign and domestic. B16
_____, Annual supplement, 1- . D3
Union list of serials in libraries of the United States and
   Canada. B17
United Nations. Dag Hammarskjold Library. Checklist of
   United Nations documents, 1946-1959. E12
_____. _____. Document Index Unit. United Nations docu-
   ments index. E12
_____. Department of Economic and Social Affairs. Popu-
   lation Division. Determinants and consequences of popu-
   lation trends; a summary of the findings of studies on the
   relationships between population changes and economic and
   social conditions. H203
_____. _____. World economic survey. F27
_____. Economic Commission for Europe. Techniques for
   surveying a country's housing situation, including estimat-
   ing of current and future housing requirements. H197
_____. Economic and Social Council. Economic Commission
   for Africa. Economic bulletin for Africa. F30
_____. _____. Economic Commission for Asia and the Far
   East. F31
_____. _____. Economic Commission for Europe. Economic
   survey of Europe. F28
_____. _____. Economic Commission for Latin America.
   Economic survey of Latin America. F29
_____. _____. International bibliography of economics. H48
_____. _____. International bibliography of political science.
   H50
_____. Food and Agriculture Organization. Catalogue of
   maps. H79
_____. _____. Trade yearbook; Annuaire du commerce;
   Annuario de comercio, 1958- . F24
_____. Headquarters Library, Bibliography on industrializa-
   tion in under-developed countries. H29

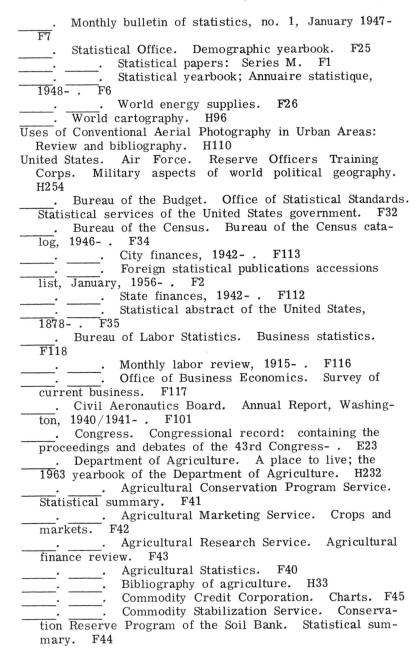

_____. Monthly bulletin of statistics, no. 1, January 1947-
F7
_____. Statistical Office. Demographic yearbook.   F25
_____. _____. Statistical papers: Series M.   F1
_____. _____. Statistical yearbook; Annuaire statistique,
1948- . F6
_____. _____. World energy supplies.   F26
_____. World cartography. H96
Uses of Conventional Aerial Photography in Urban Areas:
Review and bibliography.   H110
United States. Air Force. Reserve Officers Training
Corps. Military aspects of world political geography.
H254
_____. Bureau of the Budget. Office of Statistical Standards.
Statistical services of the United States government.   F32
_____. Bureau of the Census. Bureau of the Census cata-
log, 1946- .   F34
_____. _____. City finances, 1942- .   F113
_____. _____. Foreign statistical publications accessions
list, January, 1956- .   F2
_____. _____. State finances, 1942- .   F112
_____. _____. Statistical abstract of the United States,
1878- .   F35
_____. Bureau of Labor Statistics. Business statistics.
F118
_____. _____. Monthly labor review, 1915- .   F116
_____. _____. Office of Business Economics. Survey of
current business.   F117
_____. Civil Aeronautics Board. Annual Report, Washing-
ton, 1940/1941- .   F101
_____. Congress. Congressional record: containing the
proceedings and debates of the 43rd Congress- .   E23
_____. Department of Agriculture. A place to live; the
1963 yearbook of the Department of Agriculture.   H232
_____. _____. Agricultural Conservation Program Service.
Statistical summary.   F41
_____. _____. Agricultural Marketing Service. Crops and
markets.   F42
_____. _____. Agricultural Research Service. Agricultural
finance review.   F43
_____. _____. Agricultural Statistics.   F40
_____. _____. Bibliography of agriculture.   H33
_____. _____. Commodity Credit Corporation. Charts. F45
_____. _____. Commodity Stabilization Service. Conserva-
tion Reserve Program of the Soil Bank. Statistical sum-
mary.   F44

_____. _____. Farmer Cooperative Service. Statistics of farmer cooperatives. F46

_____. _____. Federal Extension Service. Extension activities and accomplishments. F47

_____. _____. Foreign Agricultural Service. Foreign agricultural trade of the United States. F48

United States. Department of Agriculture. Forest Service. Forest fire statistics. F50

_____. _____. _____. Tree planters notes annual edition. F49

_____. _____. Index to publications of the United States Department of Agriculture. H32

_____. _____. Land; the 1958 yearbook of agriculture. H233

_____. _____. Rural Electrification Administration. Annual statistical report. F51

_____. _____. Soil Conservation Service. Soil conservation districts. F52

_____. _____. Soil: the 1957 yearbook of agriculture. H187

_____. _____. Yearbooks of agriculture. H221

_____. Department of Commerce. Bibliographies on climate. H16

_____. _____. Bureau of the Census. Annual trade report. F55

_____. _____. _____. Census of agriculture. F54

_____. _____. _____. Census of business. F56

_____. _____. _____. Census of housing. F58

_____. _____. _____. Census of manufacturers. F59

_____. _____. _____. Census of mineral industries. F60

_____. _____. _____. Census of population. F61

_____. _____. _____. Quarterly summary of foreign commerce of the United States. F57

_____. _____. Bureau of Foreign Commerce. World Trade Information Service. Statistical series. F62

_____. _____. Bureau of Public Roads. Highway statistics: summary to 1955. F65

_____. _____. _____. Highway statistics. F66

_____. _____. _____. Highways: current literature. H37

_____. _____. Maritime Administration. Annual report of the federal maritime board and maritime administration. F64

_____. _____. _____. Handbook of merchant shipping statistics through 1958 (1949-1958). F63

_____. _____. Office of Business Economics. Survey of current business. F53

_____. _____. Weather Bureau. Bibliographies of climatic maps. H15

_____. _____. Weather Bureau (ESSA). Climatological data,
national summary. F67
_____. Department of Defense. Annual report. F68
_____. _____. Corps of Engineers. Waterborne commerce
of the United States. F69
_____. _____. Office of the Quartermaster General. Sta-
tistical yearbook of the Quartermaster Corps. F70
United States. Department of Health, Education and Welfare.
Bureau of Old-Age and Survivors Insurance. Handbook of
old age and survivors insurance statistics. F78
_____. _____. Bureau of Public Assistance. Public assist-
ance, annual statistical data for 1966- . F79
_____. _____. Children's Bureau. Statistical series. F77
_____. _____. Health, education and welfare trends. F71
_____. _____. National Office of Vital Statistics. Annual
summary of vital statistics. F75
_____. _____. Office of Education. Biennial survey of edu-
cation in the United States. F72
_____. _____. _____. Statistics of land-grant colleges and
universities. F73
_____. _____. Office of Vocational Rehabilitation. Annual
caseload statistics of state rehabilitation agencies. F80
_____. _____. Public Health Service. Health statistics
from the United States National Health Survey. F74
_____. _____. Social Security Administration. Annual sta-
tistical supplement, social security bulletin. F76
_____. Department of the Interior. Bureau of Land Manage-
ment. Public Land Statistics. F83
_____. _____. Bureau of Mines. Minerals yearbook. F84
_____. _____. Bureau of Reclamation. Crop report and re-
lated data. F86
_____. _____. Fish and Wildlife service. Fishery Statistics
of the United States. F81
_____. _____. Geological Survey. Water supply papers.
F82
_____. _____. National Park Service. State Park service.
F85
_____. Department of Justice. Bureau of Prisons. Federal
prisons, statistical tables. F89
_____. _____. Federal Bureau of Investigation. Uniform
crime reports of the United States. F87
_____. _____. Immigration and Naturalization Service. An-
nual report. F88
_____. Department of Labor. Bureau of Employee's Com-
pensation. Federal work injuries sustained during calendar
year. F90
_____. _____. Bureau of Labor Statistics. Monthly labor re-
view. F91

_____. _____. Consumer price index. F92
_____. Department of State. Operations report, Agency for International Development. F94
_____. Department of Treasury. Annual report of the secretary of the treasury on the state of the finances. F95
_____. _____. Bureau of Accounts. Combined statement of receipts, expenditures and balances of the United States Government. F97
United States. Department of Treasury. Bureau of Customs. Merchant marine statistics Washington, 1924- . F98
_____. _____. Bureau of the Mint. Annual report. F100
_____. _____. Internal Revenue Service. Statistics of Income. F99
_____. _____. Office of the Comptroller of the Currency. Annual report. F96
_____. Federal Aviation Agency. Statistical handbook of civil aviation. F103
_____. Federal Communications Commission. Statistics of the communications industry in the United States. F104
_____. Farm Credit Administration. Annual Report. F102
_____. Federal Power Commission. Electric power statistics. F105
_____. Federal Reserve System. Banking and monetary statistics. F106
_____. Federal Works Agency. Work Projects Administration. Catalogue, WPA writers' program publications; The American Guide Series, The American Life Series. E29
_____. Geological Survey. Abstracts of North American geology. H125
_____. _____. H118, H120, H121, H122, H123, H124
United States government organization manual. E25
United States. Government Printing Office. Division of Public Documents. List of classes of United States government publications available for selection by depository libraries. E45
_____. Interstate Commerce Commission. Annual report. F107
_____. _____. Transport statistics in the United States. F108
_____. Library of Congress. African Section. Africa south of the Sahara: a selected, annotated list of writings. H66
_____. _____. _____. Official publications of British East Africa. E4
_____. _____. A catalog of books represented by Library of Congress printed cards, issued to July 31, 1942. B4
_____. _____. A guide to the microfilm collection of early state records: supplement. E32

_____ . _____ . A guide to the official publications of the other American republics. E9

_____ . _____ . A guide to the study of the United States of America; representative books reflecting the development of American life and thought. H59

United States. Library of Congress. Catalog Division. List of American doctoral dissertations printed in 1912-1938. G1

_____ . _____ . Census Library Project. Catalog of United States census publications, 1790-1945. F33

_____ . _____ . _____ . General censuses and vital statistics in the Americas. F18

_____ . _____ . _____ . National censuses and vital statistics in Europe, 1918-1939; an annotated bibliography. F12

_____ . _____ . _____ . National censuses and vital statistics in Europe, 1940-1948 Supplement. F13

_____ . _____ . Copyright Office. Catalog of copyright entries: third series; part 6: Maps and atlases. H97

_____ . _____ . General Reference and Bibliography Division. Current national bibliographies, comp. by Helen F. Conover. C14

_____ . _____ . _____ . Official publications of French West Africa, 1946-1958. E7

_____ . _____ . _____ . Union lists of serials: a bibliography. D4

_____ . _____ . Library of Congress and National Union Catalog author lists, 1942-1962: A master cumulation. C2

_____ . _____ . Library of Congress author catalog: a cumulative list of works represented by Library of Congress printed cards, 1948-1952. B6

_____ . _____ . Library of Congress author catalog: a cumulative author list representing Library of Congress printed cards and titles reported by other American libraries, 1958-1962. B8

_____ . _____ . Library of Congress catalogue: a cumulative list of works represented by Library of Congress printed cards: Maps and Atlases. H80

_____ . _____ . Library of Congress catalog. Books: subjects, 1950-1954; a cumulative list of works represented by Library of Congress printed cards. B12

_____ . _____ . _____ . Books: subjects, 1955-1959; a cumulative list of works represented by Library of Congress printed cards. B13

_____ . _____ . Madagascar and adjacent islands; a guide to official publications. E6

_____ . _____ . Map Division. A guide to historical cartography, a selected, annotated list of references on the history

of maps and map making, comp. by Walter W. Ristow
and Clara E. LeGear. H81
_____. _____. _____. A list of geographical atlases in the
Library of Congress, with bibliographical notes. H82
_____. _____. _____. Aviation cartography: a historico-
bibliographic study of aeronautical charts, by Walter W.
Ristow. H84
United States. Library of Congress. Map Division. United
States Atlases: a list of national, state, county, city and
regional atlases in the Library of Congress, comp. by
Clara E. LeGear. H83
_____. _____. Official publications of French Equatorial
Africa, French Cameroons, and Togo, 1946-1958. E5
_____. _____. Photoduplication Service. A guide to the
microfilm collection of early state records. E31
_____. _____. Processing Department. Monthly check-list
of state publications, 1 vol. E36
_____. _____. Reference Department. Statistical bulletins;
an annotated bibliography of the general statistical bulle-
tins of major political subdivisions of the world. F8
_____. _____. _____. Soviet geography, a bibliography. B60
_____. _____. _____. Statistical yearbooks; an annotated
bibliography of the general statistical yearbooks of major
political subdivisions of the world. F9
_____. _____. Serials Division. Popular names of United
States government reports; a catalog. E24
_____. _____. State censuses; an annotated bibliography of
censuses of population taken after the year 1790 by states
and territories of the United States. F111
_____. _____. Subject Cataloging Division. Classification
class G; geography, anthropology, folklore, manners and
customs, recreation. A21
_____. _____. Supplement: Cards issued August 1, 1942-
December 31, 1947. B5
_____. _____. The National Union Catalog: a cumulative
author list representing Library of Congress printed cards
and titles reported by other American libraries, 1953-
1957. B7
_____. _____. The National Union Catalog: a cumulative
author list representing Library of Congress printed cards
and titles reported by other American libraries, 1963-
1967. B9
_____. _____. The National Union Catalog: a cumulative
author list representing Library of Congress printed cards
and titles reported by other American libraries, 1968. B10
_____. _____. The National Union Catalog: a cumulative
author list representing Library of Congress printed cards

and titles reported by other American libraries, 1968.
B11
_____. _____. The National Union Catalog; Pre-1956 imprints.   C1
_____. National Archives.   Guide to the records in the National Archives.   F123
United States.   National Archives.   List of National Archives microfilm publications, 1965.   F125
_____. _____. National Archives accessions, 1947- .   F124
_____. _____. Publications of the National Archives and Records Service.   F126
_____. _____. Your government's records in the National Archives.   F127
_____. National Science Foundation.   Office of Antarctic Programs.   Antarctic bibliography.   H73
_____. National Vital Statistics Division.   County and city data book, 1949- .   F37
_____. _____. Historical statistics of the United States, colonial times to 1957.   F38
_____. _____. Historical statistics of the United States, colonial times to 1957; continuation to 1962 and revisions. F39
_____. _____. Vital statistics of the United States, 1937- . F36
_____. Navy.   Oceanographic Office.   DoD Nautical Chart Library.   Accession list of domestic and foreign charts, cumulative list.   H98
_____. Post Office Department.   Annual report.   F93
_____. Superintendent of Documents.   Catalog of the public documents of Congress and of all departments of the government of the United States for the period March 4, 1893-December 31, 1940.   E20
_____. _____. Checklist of United States public documents, 1789-1909.   E19
_____. _____. Monthly catalog of United States government publications.   E21
_____. _____. Selected list of United States government publications.   E22
_____. Selective Service System.   Annual report.   F109
_____. Veterans Administration.   Statistical summary of VA activities.   F110
_____. Weather Bureau.   Selective guide to published climatic data sources.   B35

Vance, R. B.   H545
Van Cleef, E.   H546, H547, H548, H549, H550
Valvanis, S.   H551

Van Royen, William. Atlas of the world's resources. H222
Vance, R. B. H446
Vertical file index: subject and title index to selected pamphlet material. C11
Vinge, Clarence L., and Vinge, A. G. United States government publications for research and teaching geography and related social and natural sciences. B19
Visher, S. S. H552, H553
Von Arx, William S. Introduction to physical oceanography. H178

Wagley, Charles (ed.). Social science research on Latin America. H272
Wagner, Philip L. H554, H555
Walford, A. H. (ed.). Guide to reference materials.
Walters, Robert L. Radar bibliography for geoscientists. B67
Wang, Jen Yu, and Barger, Gerald L. (eds. and comps.). Bibliography of agricultural meteorology. B49
Ward, Dederick C. Bibliography of theses in geology. H134
_____. H133
_____. Geologic reference sources. B34
Warntz, W. H556
Wasserman, Paul, Georgi, Charlotte and Allen, Eleanor. Statistics sources: a subject guide to data on industrial, business, social, educational, financial and other topics for the United States and selected foreign countries. F119
Water Resources Council. Hydrology and Sedimentation Committees. Annotated bibliography on hydrology and sedimentation, 1963-1965, United States and Canada. B36
Watson, J. W. H557
Wauchope, Robert (ed.). Handbook of Middle American Indians. H271
Weaver, John C., and Lukerman, Fred K. World resources statistics, a geographic sourcebook. H7
Webb, Martyn J. H558
Weigert, Hans W. (and others). Principles of political geography. H250
_____, and Stefansson, Viljhalmur (eds.). New compass of the world; a symposium on political geography. H249
Wendt, Paul F. Housing policy: the search for solutions. H198
West Africa annual, 1962- . F19
West Indies and Caribbean yearbook, 1926/27- . F15
Westermann's Geographische Bibliographie. H148
Whitaker, J. R. H559

168 Library Research in Geography

Whitbeck, H. H560
White, Charles L. World economic geography. H215
White, C. Langdon, Foscue, Edwin, and McKnight, Tom.
Regional geography of Anglo-America. H266
Whittlesey, Derwent. Dissertations in geography accepted by
universities in the United States for the degree of Ph.D.
as of May, 1935. B29
_____. H561, H562, H563, H564
Wilcox, Jerome K. Manual on the use of state publications.
B23
Wilson, Fern L. Index of publications by University bureaus
of business research. H28
Wilson, Logan. A32
Winchell, Constance M. Guide to reference works. A7
Winings, J. W. A23
Wisconsin University. Land Tenure Center. Accessions
lists. H44
Wolfe, Roy I. and Hickok, Beverly. An annotated bibliogra-
phy of the geography of transportation. H38
Wolpert, J. H565
Wood, W. F. H566
Wooldridge, Sidney W., and East, William G. The spirit
and purpose of geography. H330
World almanac, and book of facts, 1869- . F120
Woytinsky, W. S. Die welt in zahlen. H8
_____. World population and production, trends and outlook.
H9
Wright, J. K. H567, H568, H569, H570, H571, H572
_____, and Platt, Elizabeth T. Aids to geographical re-
search; bibliographies, periodicals, atlases, gazetteers
and other reference books. B1
Wyer, J. R., Jr. E56
Wynar, Bohdan S. Introduction to bibliography and reference
work; a guide to materials and sources. A6

Yearbook and guide to East Africa, 1950- . F20
Yeates, Maurice H. An introduction to quantitative analysis
in economic geography. H213
Yonge, Ena L. H103, H104, H105
Younger, R. M. The changing world of Australia. H296

Zakrzewska, B. H573
Zelinsky, Wilbur. A bibliographic guide to population geog-
raphy. B47
_____. A prologue to population geography. H204
Zipf, G. K. H574
Zobler, L. H575
Zoological Society of London. Zoological record. H138